Interconnections between Culture and Sustainability

Anna Maria Biedermann
Francisco Javier Galán Pérez
Editors

Interconnections between Culture and Sustainability

palgrave
macmillan

Editors
Anna Maria Biedermann
Department of Design and Manufacturing Engineering
University of Zaragoza
Zaragoza, Spain

OAAEP Research Group
Funded by the Government of Aragón with FEDER Funds
Zaragoza, Spain

Francisco Javier Galán Pérez
Predepartmental Unit of Fine Arts
University of Zaragoza
Zaragoza, Spain

OAAEP Research Group
Funded by the Government of Aragón with FEDER Funds
Zaragoza, Spain

ISBN 978-3-032-21060-9 ISBN 978-3-032-21061-6 (eBook)
https://doi.org/10.1007/978-3-032-21061-6

This Palgrave Macmillan imprint is published by the registered company Springer Nature Switzerland AG.
The registered company address is: Gewerbestrasse 11, 6330 Cham, Switzerland

Contents

Notes on Contributors

Laura Asión-Suñer is assistant professor in the Department of Design and Manufacturing Engineering at the University of Zaragoza. She holds a PhD in Design and Manufacturing Engineering (2022) focused on modular design and its applicability to prosumer users, exploring how modular and user-centered design methods can empower users to actively participate in product creation and customization. Her interdisciplinary research integrates sustainable design, Product-Service Systems, prosumer theory, and cultural impact assessment. She has published in high-impact journals, including *Journal of Engineering Design, Sustainability, and Biomimetics*, and has contributed extensively to international conferences and book chapters.

Anna Maria Biedermann is an associate professor in the Department of Design and Manufacturing Engineering, University of Zaragoza, Spain. Her research focuses on sustainable design, life-cycle sustainability assessment, and product–service systems, with multiple peer-reviewed publications in international journals. She investigates methodologies for integrating environmental, economic, social, and cultural dimensions in complex systems and cultural services design. Biedermann also works as a designer of travelling exhibitions. Her work has been presented in cities including New York, Los Angeles, Toronto, Quebec, Brisbane, Sydney, Cape Town, Madrid, London, and Paris. Her exhibition designs integrate research findings on sustainability and visitor experience.

Teresa Blanco-Bascuas holds a PhD in Design, Manufacturing, and Industrial Project Management and has transdisciplinary predoctoral and postdoctoral training in the humanities, design engineering, and innovation. She is a lecturer in the area of Engineering Graphic Expression at the University of Zaragoza and a researcher at the Aragon Institute of Engineering Research (I3A), where she is a member of the Human OpenWare Research Group (HOWLab). She has participated in numerous research and technology transfer projects, leading several of them, and has promoted two university spin-offs, including initiatives in the field of arts and technology. Her current research focuses on systemic and hybrid design approaches in onelife ecosystems, aimed at improving human experiences and quality of life in complex environments.

Laura Diago Ferrer is Architect and Interim Lecturer of Graphic Expression and Digital Design at the School of Engineering and Architecture, University of Zaragoza. She holds a Master's in Product Design and a degree in Architecture with a specialization in Urbanism, and manages the architectural practice *Diago Arquitectos* since 2004. At UNIZAR she teaches and researches sustainable design, eco-design education, and design methods that integrate circular economy principles into engineering curricula. Her scholarly work includes chapters on reflective design practice and innovation in design education, contributing to sustainability learning and design theory.

Estrella Sanz Domínguez is assistant professor in the Department of Painting and Conservation-Restoration at the Universidad Complutense de Madrid, where she teaches and coordinates courses on preventive conservation and heritage protection. She holds a Doctorate in Fine Arts with a thesis on textile collections in military museums and a Magíster in Preventive Conservation of Art and Cultural Property. Her research and professional practice focus on risk management, preventive conservation, and emergency protection of cultural heritage collections, with work for institutions including the Museo del Prado and Spanish public museums. Sanz Domínguez contributes to sustainable museography and collection protection methodology, and her publications cover conservation strategies and risk mitigation in heritage contexts.

Aranzazu Fernández-Vázquez is architect and lecturer in the Department of Engineering Design and Manufacturing at the University of Zaragoza, where she teaches graphic expression and spatial design. She currently works for the Municipal Housing Company of Zaragoza City

Council developing European and urban renewal projects through building retrofitting while continuing her academic activity. She has extensive experience in the design and construction of retail spaces integrating spatial quality, functionality, and user experience. Fernández-Vázquez is currently undertaking her doctoral research, focused on the analysis of sustainability in retail spaces, exploring design strategies to improve their environmental performance and long-term viability combining architecture-design-engineering perspectives.

Francisco Javier Galán Pérez is a scholar and artist at the Universidad de Zaragoza, affiliated with the Observatorio Aragonés de Arte en la Esfera Pública and the Faculty of Social and Human Sciences. He holds a PhD in Art History, a Master's in Digital Arts, and a degree in Fine Arts. His research focuses on art and technology in public space, sustainability in art practices and exhibitions, and digital artistic creation. Galán Pérez has published extensively on media façades, cultural sustainability, and exhibition design, and curated international exhibitions while producing research-led projects integrating art, technology, and social critique.

Ana Galán-Pérez is assistant professor in the Department of Painting and Conservation-Restoration at the Complutense University of Madrid, Faculty of Fine Arts. She holds a PhD in Fine Arts from the University of Seville and degrees in Conservation-Restoration and Art History. Her research focuses on preventive conservation, risk management in heritage collections, sustainability, and critical theories of conservation. She has published on sociological and sustainability perspectives in cultural heritage and contributed to international projects on conservation practice and risk in travelling exhibitions. Galán-Pérez also works professionally in collection management and preventive conservation.

Giordana Gallina is a researcher in the Area of Graphic Expression in Engineering in the Department of Engineering Design and Manufacturing, School of Engineering and Architecture, University of Zaragoza, where she contributes to design and representation education. She is currently pursuing doctoral studies focused on design research at the intersection of engineering graphic expression, participatory methodologies, and cultural processes. Specialised in International Cooperation, her work explores the role of design in community engagement and cultural contexts. She collaborates on research and applied projects linking design, culture, and territory, including initiatives within OCREA—Observatory of Cultural Creation in Rural Environments.

Guillermo Juberías Gracia is lecturer in the Department of History of Art and Musicology at the University of Oviedo, Spain. He earned his PhD in Art History with *Cum Laude* and international distinction, and previously held positions at the University of Zaragoza, Université Bretagne Sud, and Université de Lyon. His research focuses on nineteenth-century Spanish visual culture, including genre painting, the reception of Goya, and artistic exchanges between Spain and France. Juberías has participated in funded research projects and international conferences, and his work appears in peer-reviewed journals and books on contemporary art history. He is currently a member of the EsArt research group at the University of Oviedo and serves as a scientific collaborator with the Observatorio Aragonés de Arte en la Esfera Pública (OAAEP) team, funded by the Government of Aragón.

Stefano Magnolo is Assistant Professor of Sociology of Law and Sociology of Change at the University of Salento, Lecce, Italy, affiliated with the Department of Legal Sciences. He coordinates the research unit "Culture as a Good and as a Medium," which examines cultural heritage, intangible heritage, and memory within socio-legal and socio-cultural frameworks. His research spans law, science and technology, scientific communication, art and science, media, knowledge, and intangible cultural heritage, including sociological interpretations of memory and heritage. Magnolo's work has appeared in peer-reviewed journals and collaborative research on intangible cultural heritage and scientific heritage.

Eloy Bermejo Malumbres is an assistant professor in the Department of Specific Didactics at the University of Zaragoza. He develops his research activity within the ARGOS group, funded by the Government of Aragon. His current lines of research focus on the teaching of historical thinking, the application of active methodologies and the integration of artificial intelligence in teaching processes, actively participating in research and innovation projects at national and international level.

Eduardo José Manchado Pérez is associate professor in the Department of Engineering Design and Manufacturing at the School of Engineering and Architecture, University of Zaragoza, where he teaches design-related subjects including product communication, design context, and practical design workshops. He holds a PhD in Engineering Design and Manufacturing, a BA in Fine Arts (Complutense University of Madrid), and a BA Hons in 3D Product Design (De Montfort University). He is coordinator of the Master's in Product Design Engineering, and his research focuses on emotional design, product-service systems, social design, and design methodology, with publications on design education and product concept generation. His work integrates industrial, graphical, and service design approaches.

José Prieto Martin is a full professor in the Department of Fine Arts (Sculpture) at the University of Zaragoza and a researcher at the Instituto de Patrimonio y Humanidades (IPH). His work spans public art, cultural heritage, land art, and participatory practices, with exhibitions and projects across Europe. He leads research on art's role in social memory, sustainability, and community engagement, and has published extensively on heritage, collaborative art, and environmental themes. Prieto Martín also serves as Director for Cultural Action Intercampus at the university. Coordinator of the Ibero-American Network for Research in Art, Memory, and Environmental Sustainability [ArteM(y)SA].

Natalia Muñoz López is assistant professor in the Department of Design and Manufacturing Engineering at the University of Zaragoza, where she teaches graphic expression and digital design and contributes to sustainable design education. She earned her PhD in Design and Manufacturing Engineering in 2020 with a *Cum Laude* thesis on sustainability assessment of product–service systems, integrating life-cycle approaches across environmental, social, and economic dimensions. Her research focuses on sustainable design methodologies, life-cycle sustainability assessment (LCSA), and the sustainable planning of cultural services and exhibitions. She has co-authored peer-reviewed articles in *Sustainability* and *International Journal of Life Cycle Assessment.*

Carlos Romero Piqueras is associate professor in the Department of Engineering Design and Manufacturing at the School of Engineering and Architecture, University of Zaragoza, where he teaches service design, corporate image, and design projects within industrial design curricula. He has a background in industrial design and user-experience design and is a doctoral candidate in the Department of Design and Manufacturing Engineering at UNIZAR. Romero Piqueras is co-founder of Fractal Strategy, a Zaragoza-based design consultancy specialising in participatory and human-centred design processes for public institutions and organisations, including strategic service design and UX research for government and health-system projects. His academic work includes service design research in public and healthcare contexts.

Mark Lorenzo Robinson has over 20 years experience in the museum and university sectors in both Spain and the UK. Mark from the start of his career in archaeology to his current role as impact officer at the University of East London has always an interest in how cultural heritage is actively produced in the present and how it is valued and used by different audiences.

Ismael Piazuelo Rodríguez is assistant professor at the University of Zaragoza. He holds a PhD in Education (2024) and focuses his research on interculturality and inclusion, curriculum development, the teaching of historical memory, and the educational uses of videogames. He is a member of a regional research group and collaborates with an environmental sciences institute, contributing to academic outputs and conference dissemination. Alongside his university career, he has extensive experience as a secondary-school teacher, having taught for years (2016–2023), which informs his research with strong classroom-based and applied perspectives.

José Luis Santolaya Sáenz is a professor in the Department of Design and Manufacturing Engineering at the School of Engineering and Architecture, University of Zaragoza, where he specializes in sustainable design, product-service systems, and engineering graphic expression. He holds a Doctorate in Industrial Engineering and leads research on methodologies that integrate environmental, economic, and social criteria into sustainable product and service design. His work includes contributions on life-cycle sustainability assessment and sustainable cultural exhibition planning, published in international journals and co-authored with colleagues in multidimensional sustainability research. Santolaya Sáenz also supervises theses and teaches design and graphics courses.

Rosana Sanz-Segura is assistant professor in the Department of Engineering Design and Manufacturing at the School of Engineering and Architecture, University of Zaragoza, where she teaches design methodology. She holds a PhD in Design and Manufacturing Engineering and researches design processes and multidisciplinary frameworks, with publications in International Journal of Design and Journal of Service Theory and Practice among others. Her work focuses on sound-driven design and methodologies applied to complex contexts. She leads the R&D&I activities of the association OCRE and is founder of OCREA, a research observatory supporting cultural creation in rural contexts through university–community collaboration.

Fabiane Cristina Silva Dos Santos is a Professor of Fine Arts at the University of Zaragoza and a cultural researcher. With a PhD in Visual and Intermedia Arts from the Polytechnic University of Valencia, her practice focuses on gender, public space, cultural heritage, collaborative art, and new media. She integrates physical and virtual spaces in artistic narratives and contributes to research on participatory art and visual culture. Silva Dos Santos coordinates academic programs, leads creative projects, and participates in international exhibitions and publications on contemporary art.

Eva Tausiet is a researcher associated with the Human OpenWare Research Lab (HOWLab) at the Aragón Institute for Engineering Research (I3A), University of Zaragoza, where she contributes to multidisciplinary work on human-centred technology and connected systems. She has co-authored peer-reviewed research on integrated care ecosystems and assistive wearable technologies, including contributions to the Lecture Notes in Mechanical Engineering series, demonstrating her engagement with IoT-enabled health and ambient assisted living technologies. Her work reflects an interest in wearable systems, digital health integration, and design methodology for technologies that improve quality of life, and she is currently pursuing her doctoral studies in related fields within the I3A research environment.

List of Figures

List of Tables

CHAPTER 1

Introduction

Anna Maria Biedermann,
and Francisco Javier Galán Pérez

The influence of culture on our lives is an undeniable, fundamental force. Yet, much like our own breathing, we often don't realize it takes place, silently operating in the background, sustaining the very framework of our identity and worldview. It is something we rarely pause to reflect on, not because it lacks relevance, but precisely because it is an omnipresent, constant presence that shapes us as individuals and as a society. We

A. M. Biedermann (✉)
Department of Design and Manufacturing Engineering, University of Zaragoza, Zaragoza, Spain

OAAEP Research Group, Funded by the Government of Aragón with FEDER Funds, Zaragoza, Spain
e-mail: anna@unizar.es

F. J. Galán Pérez
Predepartmental Unit of Fine Arts, University of Zaragoza, Zaragoza, Spain

OAAEP Research Group, Funded by the Government of Aragón with FEDER Funds, Zaragoza, Spain
e-mail: jgalan@unizar.es

A. M. Biedermann, F. J. Galán Pérez (eds.), *Interconnections between Culture and Sustainability*,
https://doi.org/10.1007/978-3-032-21061-6_1

overlook it precisely because it is as essential and indispensable to our existence as the air we breathe and cannot see. Culture is not merely a backdrop to human behavior; it is the medium through which we interpret meaning, construct reality, and negotiate our place in the world. It molds our thoughts, informs our emotions, and structures our institutions. Its influence extends from the intimate realms of personal belief to the sweeping dynamics of political ideologies, economic systems, and collective memory. This ever-present, ever-changing nature makes culture notoriously difficult to pin down and measure. Like the process of aging, its daily changes are imperceptible, yet its lifelong transformation is total.

The relationship between culture and sustainability is intrinsic and profound, though often underestimated in global development discourses. For the authors of this book, the conceptualization of sustainability is built upon two essential perspectives. First, we depart from the premise established by the Brundtland Commission (1987), which defines sustainability as development that meets the needs of the present without compromising the ability of future generations to meet their own needs. Building on this, we argue that ensuring true sustainability requires achieving a meaningful equilibrium between four fundamental dimensions: social, economic, environmental, and cultural. Under this framework, sustainable development efforts must focus on active management aimed at minimizing negative impacts and maximizing positive ones equitably across each of these dimensions. The concept of cultural sustainability continues to evolve, framing culture as a lens to engage with and redefine the evolving functions of sustainable development within society[1] and analyze the spatial, social and economic implications it has.[2] Culture, understood as the fabric of values, beliefs, knowledge, and practices that define a society, is not a mere spectator in the ecological crisis, but a principal actor. For numerous indigenous and rural communities the dichotomy between culture and nature often does not exist; their traditional knowledge systems, agricultural practices, and worldviews are deeply intertwined with the rhythms of their ecosystems. They do not just "protect" heritage; they live in a symbiotic relationship with their environment that is, in essence, a

[1] Katriina Soini & Inger Birkeland (2014). Exploring the scientific discourse on cultural sustainability. *Geoforum, 51*, 213–223. https://doi.org/10.1016/j.geoforum.2013.12.001

[2] Kong, L. (2009). Making Sustainable Creative/Cultural Space in Shanghai and Singapore. *Geographical Review, 99*(1), 1–22. https://doi.org/10.1111/j.1931-0846.2009.tb00415.x

model of sustainability. Even in urban societies culture is the prism through which communities interpret their environment, define their aspirations, and shape their patterns of production and consumption. Therefore, any attempt to advance toward a sustainable future that does not consider the cultural dimension is destined to be incomplete, as sustainability must be rooted in the frameworks of meaning that culture provides. For this relationship to transcend the anecdotal and be effectively integrated into policies and strategies, it is essential to address the challenge of its evaluation. We need frameworks capable of linking cultural impacts with environmental, economic, and social impacts not to "reduce" culture to numbers, but to make visible its contribution to well-being, resilience, and the health of ecosystems, allowing for a more equitable dialogue with the indicators that currently dominate decision-making.

Despite culture´s elusive quality, we cannot ignore the tangible consequences of this intangible force. Culture powerfully influences our responses to pressing issues such as climate change or public health crises, shapes political actions and consumer choices, underpins our social structures, and serves as the very glue that binds communities together. Cultural norms determine what is acceptable, desirable, or taboo, and thus play a central role in shaping policy outcomes and collective action. However, because it resists easy classification, any attempt to define it, measure its impact, or assess its value is fraught with difficulty, complexity and limitations, although the intents to do it have long history.

The concept of "cultural impact" emerged in the twentieth century, shaped by the sociological, economic, and political need to measure and justify the value of culture within an increasingly quantifiable world. Yet, the absence of the term did not imply the absence of the idea. For centuries, philosophers, poets, art critics, and statesmen had eloquently articulated the profound influence of the arts, customs, and heritage on both individuals and society. Long before the advent of impact studies and value metrics, a robust and passionate discourse addressed the ways in which culture shapes both the inner life and the external environment. Tracing this discourse uncovers the genealogy of an idea that today stands as a cornerstone of sustainable development and cultural policy.

In the early Enlightenment, while many philosophers sought universal laws of human reason, Giambattista Vico proposed a radically different idea. In his masterpiece "The New Science" (1725), Vico argued that the social and civil world is fundamentally a human creation and can only be understood through its own products like languages, myths, laws, and arts, marking one of the first systematic articulations of cultural impact. For Vico, culture was not a secondary product of society but the very engine of meaning-making, intrinsically linking the nature of government to the character of its people. This notion of culture as the connective tissue of society, shaping morality and institutions, carried through to the Enlightenment's emphasis on art and culture as tools for moral education, is seen in Denis Diderot's art criticism. The transition to Romanticism deepened this perspective through Johann Gottfried von Herder's concept of the *Volksgeist*, which is a people's unique "spirit" expressed in language, folklore, and traditions that forges national identity and collective belonging. English Romantic poets such as Percy Bysshe Shelley expanded the idea further, presenting art's impact as intrinsic, cultivating imagination and empathy as the moral foundation of a just society.

In the Victorian era, amid the upheaval of the Industrial Revolution, cultural impact became framed as an urgent social necessity. John Ruskin, an influential art critic and social reformer, saw the aesthetic health of a nation as inseparable from its moral and social wellbeing, condemning the dehumanizing effects of mass production and praising the moral superiority of craftsman ornamentation for allowing individual artistic expression. For Ruskin, art was not a luxury but a vital necessity, measurable not in economic terms but in human dignity and happiness. Matthew Arnold, in "Culture and Anarchy" (1869), positioned culture as the unifying force against the materialism, sectarianism, and arrogance fracturing Victorian society. His vision framed cultural impact as a catalyst for social progress, compelling individuals to work for the common good and countering extreme individualism and social fragmentation.

From Vico to Arnold, we see that the discourse on culture evolved along a clear trajectory: Vico's foundational idea that culture creates the social world; the Romantic vision casting culture as the soul of the nation and the source of moral empathy; and the Victorian defense by Ruskin and Arnold of culture as a societal necessity, counterbalancing industrial dehumanization. While none of these thinkers used the term "cultural impact," they collectively laid its conceptual foundations. The twentieth century's shift toward the language of "impact," "value," and "metrics" reflected the demands of a world seeking economic and statistical justification for culture, yet it risks overlooking the enduring wisdom of these predecessors. They remind us that culture's deepest impacts are not always measurable but remain fundamental, a truth that continues to inform the now-specialized field of cultural impact assessment, closely tied to the concept of sustainable development.

While the search for a final definition of "cultural impact" and its relationship with sustainability continues, a range of methods and practical applications already exist across different countries. There is no single method for measuring cultural impact; instead, practitioners use a combination of quantitative, qualitative, and mixed approaches to capture its diverse dimensions. Logical frameworks and theories of change are often the starting point. The Logic Model[3] maps the sequence from inputs and activities to outputs, short- and medium-term outcomes, and long-term impacts, while the Theory of Change[4] takes a more narrative form, working backwards from the desired impact to identify all the conditions and intermediate steps necessary for change. Quantitative approaches focus on assigning numerical values to impact, including economic analysis (e.g., employment, Gross Value Added, tourism revenue), Social Return on Investment (SROI), which monetizes social value, and contingent valuation, which estimates the value of cultural goods without a market price through willingness-to-pay surveys. Qualitative approaches aim to capture the "how" and "why" of impact through case studies, interviews, focus groups, and participant observation, providing insights that numbers cannot fully reflect. Indicator-based frameworks bring structure, such as Arts Council England's valuation framework, which measures cultural contributions to community belonging and collaboration, or UNESCO's Culture 2030 Indicators, which assess culture's contribution to the Sustainable Development Goals across four dimensions: Environment and Resilience, Prosperity and Livelihoods, Knowledge and Skills, and Inclusion and Participation. These methods are applied differently around the world. In the United Kingdom, Arts Council England requires funded organizations to use its Impact & Insight Toolkit to collect standardized audience self-assessment, and peer-review data, making impact reporting an integral part of funding cycles. South Korea has gone further, institutionalizing Cultural Impact Assessment by law since 2014, applying it

[3] McCawley, P. F. (1997). *The Logic Model for Program Planning and Evaluation.* University of Idaho Extension.

[4] Weiss, C. H., & Connell, J. P. (1995). Nothing as practical as good theory: Exploring theory-based evaluation for comprehensive community initiatives for children and families. In J. P. Connell, A. C. Kubisch, L. B. Schorr, & C. H. Weiss (Eds.), *New approaches to evaluating community initiatives: Concepts, methods, and contexts* (pp. 65–92). Washington, DC: Aspen Institute.

even to non-cultural public policies, such as urban planning, to evaluate their effect on citizens' cultural quality of life, adopting a preventive, macro-level approach. At the European Union level, the European Commission mandates impact assessments for major policy initiatives, with cultural programs like Creative Europe encouraging applicants to detail the expected impact of their projects, although rigorous requirements vary.

Cultural value is not universal; what constitutes "cultural wealth" or "cultural loss" is perceived differently across contexts and communities. Whose values do we prioritize—those of the creator, the funder, or the audience? How to measure the impact without judging the viewpoints and preserving the diversity? It often seems easier to list the difficulties of cultural evaluation than to propose a system to navigate them. As a result, the field continues to grapple with a fundamental tension: we recognize the vital role of culture in shaping society, yet struggle to assess its contributions without compromising its complexity.

We must, however, consider the alternative. What is the risk and which are the consequences of *not* talking about cultural impact? What is the risk of *not* attempting to measure it? This risk is equally real and tangible, with consequences for the funding of cultural activities, the preservation of our heritage, and the formation of educational policies. Funding bodies often rely on metrics to allocate resources, and when culture lacks visible indicators of impact, it risks being sidelined in favor of sectors with more readily quantifiable outcomes. It is difficult not to notice that we live in a system that quantifies what counts, and what cannot be defined is often overlooked or quantified in the units that don´t represent its proper significance. This has a direct effect on policymaking as well. If cultural value is not clearly communicated, it becomes difficult to defend its place in education systems, urban development plans, or international cooperation agendas, etc. Programs that foster creativity, critical thinking, and social cohesion may be cut or deprioritized simply because their benefits are not expressed in terms decision-makers understand.

It is from this conviction that this book emerges. We believe the effort to reflect on these challenges is more urgent and necessary than ever, even if we accept that success is not guaranteed. In a context where cultural policies are increasingly expected to demonstrate relevance, effectiveness, and accountability, avoiding the discussion is no longer an option. Perhaps true success lies not in finding a single, perfect definition, but in participating in the many attempts to do so without surrender.

The present book is an interdisciplinary examination of the diverse ways culture influences sustainability. We do not pretend to offer definitive answers. Rather, the authors who have contributed to these pages each offer their unique perspective—their "grain of sand"—to form a kaleidoscopic view of the relationship between culture and sustainability. Through case studies ranging from rural cultural events to urban retail spaces, the chapters demonstrate how culture can act as both a barrier and a facilitator to a sustainable future.

In the first chapter José Luis Santolaya Sáenz, Natalia Muñoz López and Laura Diago Ferrer explore various approaches that define the relationship between culture and sustainability, in relation to the dimensions traditionally understood as constituting sustainability: namely, the environmental, economic, and social dimensions. At the theoretical level, this discussion identifies three distinct yet complementary and overlapping roles for culture in sustainable development. Recognizing these roles could facilitate the collection of data and evidence better suited to evaluating both the function and the significance of culture within sustainable development.

Drawing on fifteen years of insider experience as a cultural impact evaluator, Mark Lorenzo Robinson's chapter offers a sharp critique of why impact measurement is failing within the UK's cultural sectors. He argues that current assessment methods, rooted in neoliberal and managerialist agendas, have devolved into a performative, bureaucratic exercise used for advocacy and accountability rather than genuine evaluation. Lorenzo contends that this distortion threatens the sector's long-term sustainability by promoting homogenization and short-termism, ultimately undermining the vibrancy and experimentation vital to cultural work. The chapter calls for a fundamental shift away from superficial metrics, advocating for resources to be reinvested in supporting the integrity and originality of cultural outputs themselves - the very qualities that, throughout history, have generated the most meaningful impact.

The following chapter written by Laura Asión-Suñer and Anna Maria Biedermann focuses on the analysis of cultural impact, specifically in the context of exhibition design, examined through a sustainability lens across the entire life cycle. It considers potential impacts during the creation phase, service provision, and end-of-life stage, while also taking into account the multiple stakeholders involved. This reflection enables the linking of identified cultural impacts with those typically assessed through the Life Cycle Sustainability Assessment (LCSA) methodology.

The chapter by Estrella Sanz Domínguez reviews the main conservation measures (restoration, curative conservation, and preventive conservation) while examining how sustainability and cultural impact can be integrated into cultural heritage preservation practices. It advocates placing sustainability at the core of restoration practice at every stage of the conservation process, in parallel with the application of the Heritage Impact Assessment (HIA) tool to proactively evaluate the potential consequences of interventions.

Aranzazu Fernández-Vázquez and Anna Maria Biedermann analyze retail spaces as not only as commercial venues but also as dynamic social and cultural hubs, where interactions, identities, and everyday practices unfold. By going beyond the economic dimension, they highlight how these environments contribute to shaping collective experiences, urban life, and cultural meaning, thus making them key sites for integrating a cultural perspective into sustainability analysis. This chapter examines in detail the symbolic and experiential significance of retail spaces, considering how design choices and spatial arrangements influence the ways people relate to them. In addition, it incorporates a gender perspective, foregrounding how diverse groups may experience and interpret retail environments differently, and stressing the importance of inclusivity and equity in sustainable design practices.

Teresa Blanco and Eva Tausiet's chapter analyzes modern technological ecosystems from a cultural perspective, examining how they have transformed products into network nodes and users into interconnected subjects. It argues that these digital environments actively shape cultural meanings and social norms, while also creating critical challenges like cultural homogenization and sociotechnical exclusion. Ultimately, the study advocates for new design methodologies that approach the creation of technology as an opportunity to consciously build more sustainable and culturally inclusive future.

The focus on the preliminary design phase, particularly for cultural projects in rural areas, is the subject of analysis in the third chapter written by Rosana Sanz-Segura, Carlos Romero Piqueras, Giordana Gallina, and Eduardo Manchado. This study adapts methods traditionally applied in product design and transfers them to the design of cultural services, taking into account various dimensions of sustainability. It places particular emphasis on social cohesion, as well as on collaboration between institutions and active citizen participation.

The chapter written by Stefano Magnolo and Ana Galán-Pérez is based on a case study. It explores the role of Museo della Memoria e dell'Accoglienza (Nardò, Italy) in the preservation of intangible cultural heritage and in the mediation of dissonant heritage, employing a multidimensional methodology that combines interviews, fieldwork, and European cultural sustainability indicators. It proposes a model for assessing the relational impact of memory institutions, demonstrating how cultural heritage can contribute to sustainable development and to the construction of inclusive narratives.

The authors Guillermo Juberías Gracia, Eloy Bermejo Malumbres, and Ismael Piazuelo Rodríguez examine how cultural programming can drive sustainability and social cohesion in depopulated rural areas of Spain, focusing on the village of Fuendetodos as a case study. The chapter highlights initiatives that combine art, heritage, and environmental awareness to strengthen community identity, attract visitors, and boost the local economy. The study also proposes educational and institutional strategies to position Fuendetodos as a model of rural revitalization aligned with the Sustainable Development Goals.

Francisco Javier Galán Pérez examines the massive sculpture "Splash" made for the 2008 Zaragoza International EXPO as a case study on the intersection of art, technology, and cultural sustainability. The sculpture's journey from a celebrated public installation to being dismantled and stored highlights how cultural legacies and their symbolic narratives can be disrupted. The text argues that true cultural sustainability requires not just physical preservation but continuous public engagement and narrative activation to maintain relevance for future generations.

The chapter written by José Prieto Martin and Fabiane Cristina Silva Dos Santos presents the "Trash + Fantastic Beasts" project, a collaboration where participants create sculptures from recycled materials inspired by the medieval bestiary at the city of Teruel's Cathedral (Spain). The initiative aims to foster critical reflection on consumerism and sustainability by redefining waste as a valuable cultural resource. Ultimately, the project serves as a model for how collaborative art can merge heritage interpretation with environmental education to inspire community dialogue and eco-social responsibility.

As editors, our goal has been to offer the reader a varied and thought-provoking narrative, produced by an interdisciplinary group of researchers from fields as diverse as art history, design, engineering, architecture, sociology, and heritage conservation. We argue that understanding the cultural dimension of sustainability is essential for developing holistic and effective policies. This book is therefore a debate, not a treatise. It will likely answer some questions while creating new ones that demand future exploration, much like culture itself. The fact that a destination may be unreachable does not diminish the value of the journey. For us, working on this book was a challenge, an exercise in resisting frustration, and a profound opportunity to broaden our own perspectives. We hope it will do the same for you, our reader, encouraging you to reflect on the vital role of culture within your own personal and professional life.

Finally, we wish to thank all the chapter authors for sharing their knowledge and insights with us, as well as the publisher for their faith in and support for this research.

CHAPTER 2

Approaches That Explore the Relationship Between Culture and Sustainability

José Luis Santolaya Sáenz, Natalia Muñoz López, and Laura Diago Ferrer

Introduction

The development of peoples and their culture are two realities that can hardly be understood separately. They constantly interact and culture is essential for sustainable development and sustainability is required for cultural growth. The transformation sought by the 2030 Agenda (United Nations, 2015) and the Sustainable Development Goals (SDGs) is towards

J. L. Santolaya Sáenz (✉) • L. Diago Ferrer
Department of Design and Manufacturing Engineering, University of Zaragoza, Zaragoza, Spain
e-mail: jlsanto@unizar.es; lauradf@unizar.es

N. Muñoz López
Department of Design and Manufacturing Engineering, University of Zaragoza, Zaragoza, Spain

OAAEP Group, Funded by the Government of Aragón with FEDER Funds, Zaragoza, Spain
e-mail: nataliam@unizar.es

A. M. Biedermann, F. J. Galán Pérez (eds.), *Interconnections between Culture and Sustainability*,
https://doi.org/10.1007/978-3-032-21061-6_2

a culture that recognizes the deep connection that join people to the health of the planet and that understands the redistribution necessary for all of us to equally achieve prosperity. For this transformation to happen, it is necessary to adopt certain values and lifestyles that are the essence of the SDGs: democratic alliances, solidarity, respect for diversity, equality,... Thus, feeling part of a world in which everything is interconnected and it is necessary an holistic vision where culture values the intangible and the common.

We live in cultural ecosystems at a local, national, regional, and also global scale, made up of different elements: traditions, knowledge, values, education, natural heritage and cultural legacy (tangible and intangible),.... Currently, these ecosystems are seeing their cultural diversity in danger, a process that has many similarities with the disappearance of natural ecosystems. In this context, culture should become a place of experimentation where processes of changing narratives and building a new system of values can be designed. A place in which we all recognize ourselves with a common origin and future, and we know the place we occupy in relation to the rest of the beings that inhabit the Earth.

Culture Concept

According to this sentence "Culture is one of the two or three most complicated words in the English language" (Williams, 1976), we can detect that does not respond to a simple and immediate definition. The word culture has constantly changed in meaning and connotation, making it clear that there are words that experience continuous renewal.

The Romans used the term culture to refer to the work carried out to prepare the land for cultivation and to allude to the act of redeeming cult of those gods connected to the agricultural peoples. From the Roman Empire to the sixteenth century, the meaning of culture in Europe is associated with the people's concern for agricultural production. During the Renaissance, the term culture begins to be used in reference to the exclusive formative process of artists, philosophers, writers and those who exercised power and constituted privileged groups. At that time, it was considered that only these people's group were part of the cultural process. With the arrival of the Enlightenment period, culture is used by different authors to reinforce the idea of the human being as a rational being, as the only being capable of increasing his knowledge through the use of his will and intellect (Álvarez de Miranda, 1992). At the beginning of the nineteenth century, both terms culture and civilization, were used almost interchangeably, especially in French and English languages.

In 1798, the Dictionnaire included the following definition of culture: "Culture or civilization is that complex which includes knowledge, beliefs, art, morals, law, customs and any other habits and capacities acquired by man as a member of society" (Tylor, 1871). In 1982, the United Nations Educational, Scientific and Cultural Organization declared: "... culture gives man the capacity to reflect on himself. It is culture that makes us specifically human beings, rational, critical and ethically committed. Through culture we discern values and make choices. Through culture, man expresses himself, becomes aware of himself, recognizes himself as an unfinished project, questions his own achievements, tirelessly seeks new meanings and creates works that transcend him" (UNESCO, 1982).

Recently, Worts (2004) provides a definition of culture which may be a reference point for debate and discussion: "...the sum total of all values, collective memory, history, beliefs, mythology, rituals, symbolic objects and built heritage which reflect the manner people relate to those aspects of life which: they can know and control; as well as, those they cannot fully understand or control, but to which they need to have a conscious relationship". Often, two distinct higher level distinctions are drawn, broad-based and narrowly defined (Fakhimi, 2016): "a broad, lifestyle-based concept referring to all domains of human life, which is akin to Williams' way of life, an anthropological-archaeological interpretation, and on the other side, a narrow, art-based culture referring to both the general process of intellectual and spiritual or aesthetic development and its results".

In addition, we can bring in the symbolic dimension of culture: culture as semiotic, drawing on symbols as vehicles, arguably as the broadest view of all, including as it does both intentional and unconscious behaviour. Whilst used in different ways in several distinct intellectual disciplines and distinct systems of thought, culture is additionally also an everyday concept, it has 'public' meanings and understandings, and is used in many different ways and contexts. Its meaning has changed through time as well, from early ideas of culture as action in real life-worlds and its interaction with nature, which are essential aspects for anthropological use of the concept even today, to culture as the cultivation of the human mind and behaviour.

Sustainability Concept

The concept of sustainability was first disseminated through the Our Common Future Report (1987), led by the then Norwegian Prime Minister G. H. Brundtland. This report defined sustainable development as "that meeting the needs of the present generation without compromising the ability of future generations to meet their own needs". This concept involved three dimensions or pillars: environmental, economic and social (Fig. 2.1). Comprised of three intersecting circles, this model posits that sustainability occurs when the human, environmental and economic considerations overlap in a state of balance.

The Earth Summit, held in Rio de Janeiro in 1992, consolidated these three pillars as the paradigm of sustainable development and enshrined them as a guideline for development strategies at local, national and global levels. Although those premises already pointed towards a necessary transversal view, the truth is that the main focus was placed on the ecological dimension. Humanity's incipient environmental concerns set the path. It had only been a few years before, in The Limits to Growth report (Meadows et al., 1972), that it had been recognized for the first time that the preservation of nature went hand in hand with world prosperity. After that report, the Stockholm Conference took place under the auspices of the United Nations, and 120 countries declared the need to implement international actions to preserve what was called in that moment the human environment.

More than 50 years have passed. However, the enormous pressure we are putting on nature does not stop. Many people still perceive nature as an inexhaustible super-pantry, a vast, freely accessible storehouse available

Fig. 2.1 Sustainability based on three dimensions

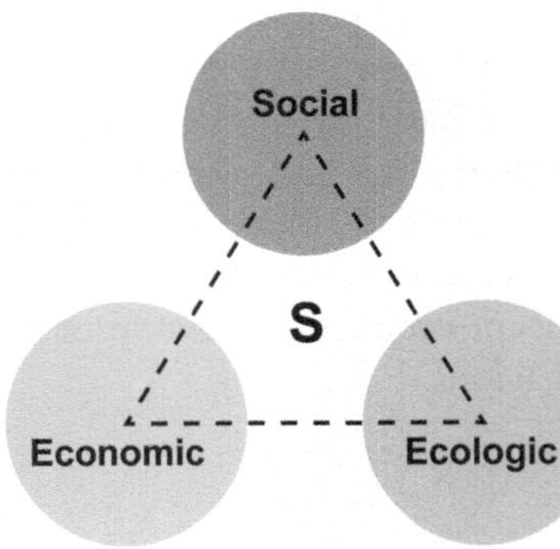

to a human population that in 1972 was made up of 3.84 billion people and today exceeds 8 billion. Our current economic system has two fundamental weaknesses: it is based on unlimited growth on a finite planet and it benefits a small number of people rather than everyone. Thus, the concept of sustainability refers not only to environmental aspects, but also to economic and social issues. It is no longer just a matter of avoid pollution or eliminating greenhouse emissions; it is also necessary to propose a balanced economic development and to combat social inequalities.

It can be noted that some changes are taking place: on the one hand, Europe is increasing environmental regulations and; on the other hand, if years ago many companies perceived sustainability as a luxury and not a necessity, today sustainability is considered a future strategy by most of them. Although it is true that there is a lot of marketing aimed at deceptively convincing us of how sustainable and green certain companies are, it is becoming increasingly difficult to lie, not only because of current legislation, but also for customers, consumers, and shareholders. Sustainability now requires the involvement of businesses and citizens, since companies cannot prosper in failing societies. Reversing injustice and environmental imbalances is required to achieve a more sustainable world.

Integration Commitment

Researchers and institutions such as UNESCO and the World Summit on Sustainable Development call for culture to be included in a sustainable development model, emphasizing that culture ultimately shapes what we understand by development and determines the way people act in the world. The world is not faced exclusively with challenges of economic, social or environmental character. Creativity, knowledge, diversity and beauty are essential prerequisites for dialogue for peace and progress, as they are intrinsically related to human development and freedom.

The SDGs cover all spheres of human activity and, therefore, each Goal can be classified in one of these dimensions or in several. The innovative vision that the 2030 Agenda proposes is the interconnection between these spheres. As a kind of communicating vessels, these dimensions respond to the aforementioned holistic and integral vision of the world and culture must commit with the SDGs for the following reasons:

- It enables the development of talent, creativity and innovation as key factors in helping the global community understand the challenges, opportunities and interactions between them.
- It helps build a critical mass of citizens with the knowledge, creativity and motivation that will help them contribute to achieving the SDGs. Passionate people, with curiosity and a desire to move towards a better world.
- It encourages working in alliance, with new partners from different disciplines, external and internal. One of the strengths of the SDGs is that they provide a common framework for different sectors and organizations to connect and work together on shared interests.
- It opens up new sources of funding. Financial institutions and sponsors, including government agencies, international banks and philanthropists, are increasingly allocating aid to the achievement of the SDGs.
- It requires us to adopt a responsible and committed attitude, more receptive and sensitive to social needs, making us agents of change.

The cultural sector has been responding to the sustainability challenge with different ideas and actions. From different institutions as the International Council of Museums (ICOM), the creation and development of sustainable solutions is promoted and on the other hand, from a more theoretical point of view, a number of studies have been carried out dealing with the integration of the culture and sustainability concepts. Several international organizations like the Council of Europe and UNESCO suggest guidelines for policy formulation aimed at integrating culture and its interconnectedness with other sustainability dimensions.

According to Worts (2006), the simple three-sphere model is a powerful tool in grasping the large concept of sustainability, but it does have some limitations. One of these is that it is hard to see how an individual can relate to this model in a practical way. It may be true that it is our collective action and reality that will determine the future of the planet, however the base unit of the collective is the individual, and we need to understand how the model affects us at that level. Secondly, there are no provisions for cultural or spiritual dimensions—considerations that have mediated human relationships with both environment and economy for thousands of years. It is through our cultural and spiritual practices that individuals and collectives have related to the large and mysterious forces that have converged to create life on the planet. If we are to have a chance of redefining our relationship with the Earth as we approach its physical and biophysical limits, we need to spend more time and energy on our cultural and spiritual perspectives. Culture demands a place in our lives beyond that of 'leisure time' and entertainment.

Throsby (2008) suggests very important guidelines for policy formulation aimed at integrating culture into sustainable development:

- Intergenerational equity: development must take a long-term view and not be such as to compromise the capacities of future generations to access cultural resources and meet their cultural needs; this requires particular concern for protecting and enhancing a nation's tangible and intangible cultural capital.
- Intragenerational equity: development must provide equity in access to cultural production, participation and enjoyment to all members of the community on a fair and non-discriminatory basis; in particular, attention must be paid to the poorest members of society to ensure that development is consistent with the objectives of poverty alleviation.
- Importance of diversity: just as sustainable development requires the protection of biodiversity, so also should account be taken of the value of cultural diversity to the processes of economic, social and cultural development.
- Precautionary principle: when facing decisions with irreversible consequences such as the destruction of cultural heritage or the extinction of valued cultural practices, a risk-averse position must be adopted.
- Interconnectedness: economic, social, cultural and environmental systems should not be seen in isolation; rather, a holistic approach is required, i.e. one that recognizes interconnectedness, particularly between economic and cultural development.

The Spanish network for sustainable development (REDS, 2021) includes the following contributions between culture and sustainability (Fig. 2.2).

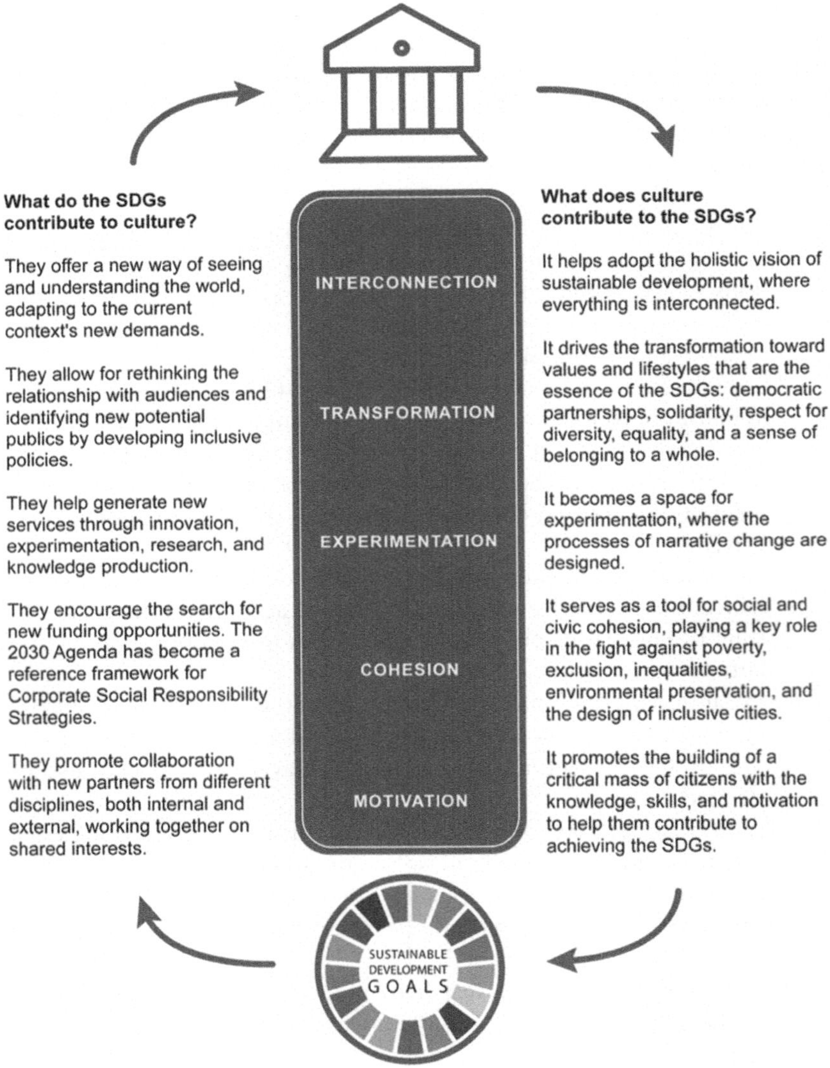

Fig. 2.2 Sustainability-culture contributions (adapted from REDS, 2021)

Culture and Sustainability: Integration Approaches

The role of culture in sustainable development is not only about "using artists to raise concern on climate change" or about "building cultural venues that are efficient in the use of natural resources". It is not only about the income that cultural industries can bring to the economy. It is not about "asking more" to the cultural circles. These are very important questions that need to be addressed, but they do not articulate the core question. It is about guaranteeing that any sustainable development process has a soul. A new perspective addresses the relationship between culture and sustainable development through a dual approach: developing the cultural sectors themselves (namely: heritage, creativity, cultural industries, art, cultural tourism); and advocating for culture to be suitably recognized in all public policies, particularly those related to education, economy, science, communication, environment, social cohesion and international cooperation. Intercultural dialogue is one of humanity's greatest challenges, while creativity is valued as an inexhaustible resource for society and the economy.

The integration of culture and sustainability concepts has been an object of study by different authors in order to obtain a comprehensive analytical framework that allows their joint and structured study. Dessein et al. (2015) have introduced three approaches that explore the ways in which culture is predominantly positioned vis-à-vis sustainability and its general role in each situation (Fig. 2.3). These approaches are based on different prevailing conceptual discourses and have been, respectively, designated as 'culture in sustainability', 'culture for sustainability', and 'culture as sustainability'.

Based on the three dimensions concept of sustainability, in the first diagram of Fig. 2.1, 'culture in sustainability', culture becomes the fourth dimension of sustainability. Cultural sustainability can be found in parallel to ecological, social, and economic sustainability, having an independent role although interconnected with all dimensions.

In the second diagram, 'culture for sustainability', culture has a mediating role to achieve all the other sustainability pillars. Thus, the culture circle connects economic, social, and ecological sustainability. Finally, the diagram 'culture as sustainability', considers culture not only as an instrument but as a necessary foundation for meeting the overall aims of sustainability. In this approach, culture encloses all other dimensions of sustainability and becomes an overarching concern or paradigm of

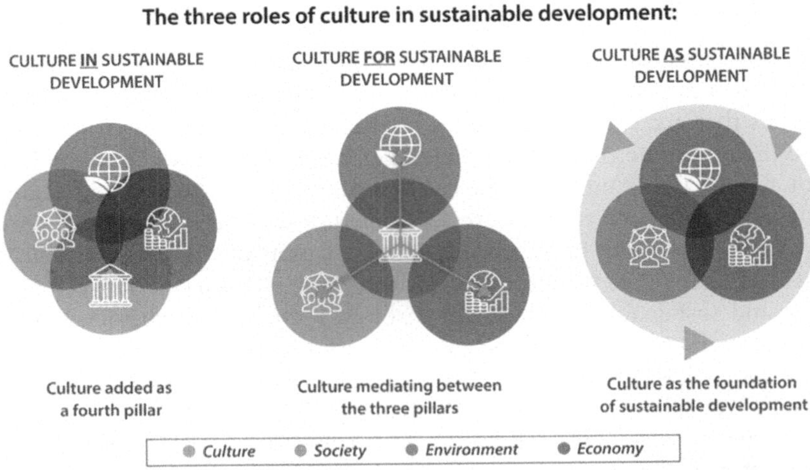

Fig. 2.3 Integration approaches of the culture and sustainability concepts (adapted from Dessein et al., 2015)

sustainability. These representations can be used to explore each general approach in more detail as well as to help in organizing and comparing different discourses. The applicability of each approach to the international cultural policy context still needs to be explored (Soini & Dessein, 2016).

Culture in Sustainable Development

The first approach, in which culture is a separate aspect, a free-standing or self-standing additional pillar, the '4th pillar', a role as an independent and autonomous dimension alongside the others, is already formulated by Hawkes (2001) who, being aware of culture's essential role in the sustainability multidimensional challenge, argues that without a foundation that expressly includes culture, the new frameworks are bereft of the means of comprehending, let alone implementing, the changes they promote. According to this author, sustainability is composed of four pillars: (1) Cultural vitality: well-being, creativity, diversity, and innovation; (2) Social equity: justice, engagement, cohesion, and welfare; (3) Environmental responsibility: ecological balance; (4) Economic viability: material prosperity.

Several transnational and international organizations like the Council of Europe and UNESCO have also advocated culture as an explicit aspect of sustainability, suggesting guidelines for policy formulation aimed at integrating culture and its interconnectedness with other sustainability dimensions. However, although holistic and cross-disciplinary integration have been proposed, this has not been done comprehensively, and the essence of culture in sustainable development research and policies therefore tends to remain ignored.

The integration based on the construction of a fourth pillar, advocates for culture to be at the same level of significance as economic, social, and environmental issues. Thus, the cultural dimension's role in sustainable development is differentiated, raising awareness of its value and generating more attention for public action. This model has been afterwards promoted by other organizations such as the United Cities and Local Governments (Agenda 21 for Culture, 2009), which intends to implement sustainability action plans and develop strategic directions based on sustainability principles to guide its own regional cultural development initiatives, as well as by authors such as Stylianou-Lamberta et al. (2014), that explore the role of culture in sustainable development and propose a theoretical model to design appropriate cultural policies for museums.

According to this model, sustainability is a balance that has to be constantly negotiated (Worts, 2004). Human development, social equity, economic productivity, and environmental quality should be simultaneously emphasized. The integration of culture as an additional sustainability dimension allows several cultural aspects can be discussed or elaborated as a different part of the social pillar, often considered as socio-cultural sustainability (Chiu, 2004). Thus, culture is recognized differently from the social dimension, so it should be separated in practice by selecting appropriate cultural sustainable parameters and applying suitable methods and tools in its evaluation. Nevertheless, it should be noted that the model considers intersected circles, in which common features are found. Culture and other dimensions have a degree of interlinkedness, in which culture can affect and be affected by environmental, economic, and social issues. Certain parameters can be common to more than one pillar.

For different authors, the argument of the fourth pillar is supported by a number of significant reasons:

- it offers a wide picture that does not instrumentalize culture;
- it preserves the intrinsic values of culture (memory, creativity, diversity, dialogue, rituality) as the core;

- it allows cultural and ecological stakeholders to connect;
- it contributes to linking culture to social actors;
- it considers cultural and creative industries as priorities;
- it empowers the cultural sector;
- it leads to the involvement of civil society in the elaboration and implementation of policies;
- it entails the growing importance of culture in international relations, cooperation and diplomacy;
- it is a clear and beautiful image.

In short, it can be said that this approach considers that cultural pillar helps us to build our human development through the explicit inclusion of cultural parameters within a connected relation with other sustainability dimensions.

Culture for Sustainable Development

In this approach, culture has a connecting and mediating role. The culture acts as driver of sustainability processes; this transcends the drawbacks and benefits of ecological, economic and social development. Economic, social and ecological sustainability afford by culture.

Since all human beings both have culture and are cultural human beings, we need a broader conceptualization of culture that includes the diversity of human values, subjective meanings, expressions and life-modes, and that allows us to distinguish between differences in culture and between cultures in a fruitful way, without making judgments about qualities of art and culture. Culture is the meaningful content of human societies and communities. It is made by individuals within societies whilst simultaneously also shaping their lives and existence. In terms of sustainability's three pillars, culture can be the way to balance competing or conflicting demands and work through communication to give human and social meaning to sustainable development. Culture can be an intermediary to connect the various dimensions of sustainability, as shown in the second part of the diagram in Fig. 2.3.

However, some sort of lens or filter is required to understand how culture mediates the relation between society and environment. One might be the concept of landscape, for example, another might be the context of territorialization, a third could be ecosystems services, and creativity might be fourth example. The fact that the potential of culture's mediating role has rarely been exploited perhaps explains why sustainable development has proved to be so elusive.

Culture as Sustainable Development

In this approach, culture takes on its evolutionary, holistic and transformative role, providing a new paradigm to the question of sustainable development.

Culture can be viewed at a more profound level of society as a core issue for a transition towards sustainable development. We can for example insist on a co-thinking of environmental, social and cultural sustainability, and an insistence on how social life is embedded in particular places and situations. A truly evolutionary culture, or an eco-cultural civilization, involves practicing a new understanding of the human place in the world, and recognizing that humans are an inseparable part of the more-than-human world. Crucially, this means that every human action is always relative to and influenced by the situation at hand. It allows new values, new ways of life, and (perhaps) utopian visions of a sustainable society.

Culture thus becomes the matrix for particular ways of life. In this sense, culture is more than a descriptive or analytical tool, and offers an ideal of doing things well, of culture as cultivation and sustaining life, but without making things well at the cost of something or somebody else. Culture in this approach refers to a worldview, a cultural system guided by intentions, motivations, ethical and moral choices, rooted in values that drive our individual and collective actions [17], and to a process and communication of transformation and cultural change. This makes it possible to think of sustainability and sustainable development as processes, ongoing and in the-making, not as fixed states.

Sustainabilities imply making connections between people and the worlds they inhabit and use. In this approach, eco-culture is deeply related to social learning by working with place-conscious and place-responsive teaching, sharing and learning, and engaging humans in discussions of what kind of world we want to live in now and in the future. This is applicable in policymaking and even in wider politics: engaging citizens in discussions of what kind of world should be a basic premise of public policies. Culture refers here not to particular types of knowledge, but to fundamental new processes of social learning that are nourishing, healing, and restorative. Sustainability exists thus as a process of community-based thinking that is pluralistic where culture represents both problem and possibility, form and process, and concerns those issues, values and means whereby a society or community may continue to exist.

Future Challenge

Sustainable development requires a reflection on the future, a prospective or forecast of subsequent effects to be taken into account. Cultural institutions do not usually use this vision, which almost never foresees or designs the future, nor does it conduct impact studies, nor is it accustomed to assessing subsequent repercussions. Culture looks to the past and already has difficulties in appreciating contemporaneity. Therefore, there are difficulties in locating the axis: past—present—future (Martinell et al., 2020). But creators and artists contribute their visions of the future and the consequences of our actions (utopias, dystopias, etc.) or anticipate social and cultural changes.

Many research works have addressed the problem of evaluating and improving the sustainability of a system. In order to achieve this objective, it is considered essential to apply a perspective that takes into account the entire life cycle of that system, to use indicators that express different impacts associated with each of the dimensions of sustainability and to use techniques that allow the analysis of the most determining impact factors and the most effective improvement strategies. In line with this objective, the design of sustainability indicators constitutes a challenge to scientists, however, given the multidimensionality and value-laden nature of sustainability, and this difficulty is only exacerbated in the context of culture which can less easily than, say economics or ecology, be quantified statistically.

A clear understanding of the linkages between culture and sustainability is required at the conceptual level. Here, our identification of three different but complementary and overlapping roles for culture in sustainable development might facilitate the collection of data and evidence that is more suitable for the evaluation of the role and meaning of culture in sustainable development. However, come questions related to the availability, standardization, aggregation, and ranking of data, all of which are required to allow cultural statistics to be consistently constructed, and therefore useful.

The different approaches proposed to integrate culture and sustainability can be understood as a progressive attempt to make a more global intervention and a more complete sustainability assessment. The approach 'culture in sustainability' can facilitate the use of those assessment methods applied so far, as long as the appropriate cultural impact indicators can be selected. The approach 'culture for sustainability' according to its mediating role, can allow the application of more balanced sustainability analysis and improvement techniques, and the approach 'culture as sustainability' can favour the involvement of all social agents (individual people, local and global communities).

The development of indicators that more usefully reflect culture should be considered as a joint learning process. This implies that the importance of incorporating cultural assessments in sustainability programmes needs to be revealed to a wider range of stakeholders, participants and researchers. In addition, good examples and practices should be urgently found and shared. What is needed, for example, is qualitative indicators, examples of different types and formats, illustrations of how they can be used, and ways for them to be combined with quantitative indicators. There already exist approaches, frameworks, and procedures which might be exploited as a starting point or work of reference when developing indicators for culture. This includes cultural ecosystem services and the principles and practice of sustainable design. But there seems to be need for many different and parallel assessment methods and types of indicators, instead of one.

In recent years, national sustainable development plans have recognized the cultural and creative sectors as a driving force for achieving results, mainly cultural, social, and economic. The inclusion of lines of action specifically related to the Convention has remained stable overall. However, too little investment is still being made in developing the cross-cutting role of culture and creativity in sustainable development, as well as in the environmental transition (this aspect being recognized in only 13% of voluntary national reviews monitoring the implementation of the 2030 Agenda). One of the main challenges for the sector is to unleash the transformative power of culture and creativity to raise awareness, act on the climate and shift towards more sustainable models. At the international level, the share of official development assistance allocated to culture and leisure activities barely reaches 0.23% (Kamara, 2022). Contributions to the International Fund for Cultural Diversity have also declined in recent years. However, the involvement of non-traditional donor countries, the growing interest of multilateral organizations in cultural and creative industries, and the rise of South-South and triangular cooperation models offer new opportunities.

References

Agenda 21 for Culture. (2009). Culture and sustainable development: examples of institutional innovation and proposal of a new cultural policy profile. http://www.cities-localgovernments.org and http://www.agenda21culture.net

Álvarez de Miranda, P. (1992). *Palabras e ideas: el léxico de la ilustración temprana en España (1680–1760)*. Real Academia Española.

Brundtland, G. H. (1987). *Report of the world commission on environment and development: Our common future*. United Nations General Assembly document A/42/427.

Chiu, R. (2004). Socio-cultural sustainability of housing: A conceptual exploration. *Housing, Theory and Society, 21*(2), 65–76.

Dessein, J., Soini, K., Fairclough, G., & Horlings, L. (2015). *Culture in, for and as sustainable development. Conclusions from the COST action IS1007 investigating cultural sustainability*. University of Jyväskylä.

Fakhimi Anbaran, F. (2016). "A whole way of life": Ontology of culture from Raymond Williams's perspective. *International Letters of Social and Humanistic Sciences, 67*, 46–56.

Hawkes, J. (2001). The fourth pillar of sustainability: Culture's essential role in public planning, cultural development network (Vic.), Common Ground.

Kamara, Y. (2022). Cultura y desarrollo sostenible: un potencial aún sin explotar. En: Repensar las políticas para la creatividad: plantear la cultura como un bien público global, 209–235. Unesco (UnesDoc).

Martinell, A. (coord.) et al. (2020). Cultura y Desarrollo Sostenible. Aportaciones al debate sobre la dimensión cultural de la Agenda 2030, REDS,.

Meadows, D. H. [and others] (1972). The limits to growth; a report for the Club of Rome's project on the predicament of mankind. Universe Books.

REDS. (2021). Hacia una cultura sostenible. In *Guía práctica para integrar la Agenda 2030 en el sector cultural*. Desarrollo Sostenible (REDS/SDSN Spain). www.reds-sdsn.es

Soini, K., & Dessein, J. (2016). Culture-sustainability relation: Towards a conceptual framework. *Sustainability, 8*, 167.

Stylianou-Lamberta, T., Boukasb, N., & Christodoulou-Yeralia, M. (2014). Museums and cultural sustainability: stakeholders, forces, and cultural policies.*International Journal of Cultural Policy, 20*(5), 566–587.

Thorsby, D. (2008, January). Culture in sustainable development: Insights for the future implementation of art. 13. UNESCO, convention on the protection and promotion of the diversity of cultural expressions.

Tylor, E.B., 1871. Primitive culture. Ed.2, J. Murray 1871.

UNESCO. (1982). Declaración de México, Cultura y civilización.

United Nations. (2015). *Transforming our world: The 2030 agenda for sustainable development*. UN. www.un.org/sustainabledevelopment/es/objetivos-de-desarrollo-sostenible

Williams, R. (1976). Keywords. A Vocabulary of Culture and Society. Routledge Revivals. Fontana/Groom Helm, London (reprinted in 2011).

Worts, D. (2004, Fall). Museums in search of a sustainable future. *Alberta Museums Review, 30*, 40–57.

Worts, D. (2006). Fostering a culture of sustainability. *Museums & Social Issues, 1*, 151–172.

CHAPTER 3

Why Impact Measuring Is Not Working: An Insider's Account from Within the Culture Sector in the United Kingdom

Mark Lorenzo Robinson

Introduction

This book chapter argues against the dominance of impact measuring and the resulting largely adverse effects on the culture sector, specifically in the museum/heritage sector and arts-based research at universities in the United Kingdom, where I work as an impact officer and evaluator. It primarily centres on my own reflections from 15 years of experience in evaluation and impact measurement within these sectors, reinforced by the insights of others both within and outside of those fields.

For me, this is a way of making sense of a career that has drifted to places I did not mean it to take, on reflections on why I felt increasingly uncomfortable with impact measuring methodologies, the results, how they were used and the effects I observed on the sector. It is also a way of

M. Lorenzo Robinson (✉)
Office for Postgraduate Research and Engagement, University of East London, London, UK
e-mail: m.lorenzo@uel.ac.uk

A. M. Biedermann, F. J. Galán Pérez (eds.), *Interconnections between Culture and Sustainability*,
https://doi.org/10.1007/978-3-032-21061-6_3

providing hope for myself in the sector, based on the idea that only by revealing the negative that is already there, by being completely honest, can we truly make progress. This chapter also looks forward to potential and alternative solutions—new approaches. For others, I hope this act of whistleblowing, so to speak, challenges and creates discussion through provocation, hopefully leading to new insights. That is my impact objective, but please, no measuring.

Impact Measuring in the University and Museum Sectors: A Short Overview

In the UK, impact measurement in higher education and the museum sector relies on formal frameworks designed to demonstrate the benefits of research and cultural activities. Universities primarily use the Research Excellence Framework (REF), which assesses research impact mainly through Impact Case Studies, brief five-page narrative-based documents supported by impact evidence in the form of testimonials, reports, statistics, and online references, collected by researchers and impact officers. Research Impact is judged on its significance and reach.

In the museum and heritage sector, impact is often measured through outcome-focused frameworks, such as those of the National Lottery Heritage Fund. Projects must demonstrate results across heritage, people, and communities, with a strong focus on social impacts, including community cohesion, social inclusion, and behavioural change. Impact measurement, gathered by project staff, evaluators, and external consultants, includes improvements to heritage, attitude shifts, participation, and learning, and is primarily collected through surveys, interviews, focus groups, and other methods. In both museums and universities, Impact measuring is at its core, an extractive process, one in which the objective is to extract from observations of human interactions/ engagement with cultural outputs—be it an exhibition or practice based arts research a series of measurable changes gathered using a range of preset quantitative and qualitative tools, The results are then presented in written form to the various funders, assessment bodies and stakeholders. The basis of the argument in this chapter is that the assumption underlying the objective that impact can be measured objectively is, in many cases, deeply flawed. These flaws are dismissed or ignored in the sector because the end product, the written report, is used for advocacy and to give the appearance of accountability, of making a case for the cultural sector in a neoliberal world.

Lastly but far from least, I argue that this whole merry-go-round is having a transformative effect on the cultural outputs themselves, be they exhibitions or a performance and jeopardises the long-term sustainability of the sector itself.

The chapter begins by exploring the similarities and differences between museum exhibitions, heritage projects and arts-based research in a university setting from an impact assessment perspective. Moving on to a discussion of the difficulties and flawed nature of measuring societal impact in these sectors, before then taking a closer look at the source for the rise of impact measuring, reporting for accountability and advocacy purposes to the various stakeholders and funders. It then examines how the obsession with measuring impact is affecting the cultural outputs themselves, before seeking to outline potential solutions to the current malaise.

The University and Museum Sectors: A Comparison

This chapter focuses on social impact measuring within the museum/heritage project sector and arts and humanities practice-based research in universities, exclusively in the UK. It is no accident that I moved seamlessly from the museum to the university sector, as they share many of the same ingredients to which my argument about impact measurement applies. The differences are primarily due to the proportion of these ingredients and the degree to which nefarious effects from impact measuring affect them.

Both sectors operate in the public realm, largely funded by public resources; museum and heritage projects are primarily supported by the National Lottery Heritage Fund, while arts and humanities research is backed through the Arts and Humanities Research Council (AHRC) via the Research Excellence Framework (REF). This shared reliance on public funding means both sectors are subject to political scrutiny and accountability, emphasising a commitment to operating in the public interest. Since the 1980s, both have experienced processes of managerialism and professionalisation that stress efficiency, performance measurement, and accountability. This transformation has altered the organisational culture of museums and universities alike.

Another similarity lies in the assumption of positive impact; both sectors believe their work contributes positively to society. They operate from centres of power and legitimacy, functioning with a pedagogical role: museums work to communicate knowledge while university research informs teaching. Through the production of new knowledge and mechanisms such as public engagement and participatory approaches, they aim to affect the world positively. Additionally, both sectors produce cultural outputs—museum exhibitions serve as cultural artefacts that disseminate knowledge and research, while arts and humanities research manifests through publications and practice-based outputs, creating new knowledge. Exhibitions interpret and communicate this knowledge, framing it as educational, sensory, and artistic rather than merely transmission vehicles.

Nonetheless, there are notable differences between the sectors, particularly in the importance given to impact measurement. For instance, in museums and heritage projects, impacting visitors is one of several aims, while for universities, impact has become a key justification for research. The scale of impact is one measure of success for museum exhibitions, alongside visitor satisfaction, audience diversity, and sales. In contrast, research impact in universities is evaluated through strict parameters, including impact case studies required for funding applications and evaluations. This competitive environment means that higher impact scores can lead to increased funding, creating incentives for over-inflating impact claims, a phenomenon I refer to as "impactflation." In museums, any perceived impact must be substantiated as part of a reporting process after funding is granted; thus, the museum sector is less prone to the hyperbole found in academic research contexts. In summary, the university sector exhibits a more intense culture of impact measurement, which can lead to negative consequences and distort the representation of true impact.

The Limitations of Impact Measurement in the Cultural Sector

You simply cannot measure the changes in behaviour and attitudes resulting from exposure to an exhibition or, for example, a performance. We can explore how they interact—through visitor studies, ethnography, observations—but fitting that into a report, or within a 500-word limited impact case study to be assessed, is a futile task and should not be done if you believe in the integrity of how results are produced.

Currently, impact is often assessed through evaluations involving verbal interactions—open or closed questions like, "How did you find the exhibition?" or "What did you learn?" and "Has your perspective changed?"

These rely heavily on verbal testimony to gauge impact. However, this approach is problematic. People tend to give positive responses, wanting to please or fit into perceived expectations. Moreover, the dynamic between evaluator and participant, the setting, mood, and timing all influence these responses.

The fundamental reason why measuring change in attitudes and learning is such a challenging, even impossible, task is that, as humans, we are active, not passive, in our interpretations of our surroundings. As Eileen Hooper-Greenhill states (Hooper-Greenhill, 2006, p. 23), "the shift from thinking about visitors as an undifferentiated mass public to beginning to accept visitors as active interpreters and performers of meaning-making practices" reflects a recognition that visitors embed and interpret content in diverse ways. This was already recognised in the 1980s, yet museums and universities still insist on measuring changes in attitudes, often viewing visitors as passive.

There has been research such as Sharon Macdonald's landmark study, "Behind the Scenes at the Science Museum" (Macdonald, 2002), which aimed to understand how visitors culturally frame their experience—how they recode and decode the exhibition. In doing so, the research sought to go beyond the typical cognitive focus on "what visitors have learned" or whether they "got" the intended messages. Ironically, and as a side note, one of the ways identified that visitors in modern museum exhibitions decode an exhibition is through consumption, which would mean that the impact on changing behaviours and learning is largely passive.

Her book illustrates the challenges and even the futility of impact measuring. Her in-depth, nuanced study of how visitors react and what they take away is a tour de force—full of detail and insight. But how do you crunch this into numbers for a report? How do you measure impact? How do you determine success? This kind of research often remains isolated, not operationalised or instrumentalised in the day-to-day running of museums or in reports. It is often ignored, with no real attempt to transfer findings into operational practice. Much of this research is rich, ethnographic, nuanced—the kind of approach that's hard to fit into the simplified, quantified formats museums tend to produce because of limited time and resources.

In light of the fact that individual backgrounds shape a person's interpretation and potential for change, it becomes even more difficult to quantify impact through metrics, which tend to overlook these differences. As numerous studies have shown—ranging from Pierre Bourdieu to Paul Willis—the value or impact of a work of art varies enormously according to factors that make up a person's identity, including age, class, health, wealth, and so on. Another way of looking at it could be to view it as the contrast between short-term engagement and long-term, sustained engagement, and how these influence a person's behaviour. For example, family, upbringing, and social class are all ongoing influences and have a huge effect on you, your way of behaving, and interpreting and decoding the world. In contrast, a short-term museum visit cannot realistically compete in terms of effecting lasting change in behaviour or outlook.

While the multifaceted nature of a person's identity complicates impact assessment, the challenge is further compounded when we consider the timescales over which genuine change unfolds and how this could ever be measured. Impact often occurs on different timescales. One of my mainstays has been the evaluation of heritage community engagement projects aimed at getting more people from underrepresented communities involved in exhibitions and heritage interpretation. These collaborative projects are potentially where impact measuring stands on firmer ground due to the close working partnership, but even here, the timescales are not commensurate with how changes are made in people's lives. Beyond stating that the action happened, and these people came, did this, and said this, we cannot be certain about the long-term impact. The practicalities of financing for the evaluator, the funding restraints, and the finite scales can only mostly describe what happened; any longer-term impacts are very challenging to measure, even with community projects. Controlling what happens with research and cultural productions in general, beyond the museum space or the university, can be extremely time-consuming. Evaluators simply don't have the resources or the freedom to carry out this task.

Take, for example, an exhibition on black dolls that I was involved in evaluating at the National Trust Museum of Childhood, whose stated impact objective was to contribute to more representation from ethnic minority visitors. Suppose we have the statistics for subsequent years and see an increase in ethnic minority visitors. How do we confidently attribute this increase to the exhibition? It could be due to numerous other factors, and ethnicity is only one aspect of identity. Moreover, studies show that identity is fluid and dependent on context. As a result, scrutinising the data reveals that such attributions are often unfounded or overly simplistic. In short, despite our desire for a clear, causal link, where the exhibition directly causes increased attendance or engagement, we simply do not have the evidence to support that certainty. Long-term impact is extremely difficult, if not impossible, to attribute definitively to a single intervention or event.

As the previous discussion has hopefully highlighted, measuring the impact of changes in individuals after experiencing a cultural output is anything but straightforward. The sector, with its associated accountability, yearns for a straightforward toolkit method of impact evaluation (Belfiore, 2009, p. 3), and in this quest, it has tended to privilege quantitative approaches largely borrowed from the disciplines of economics and auditing. Where this is particularly prevalent in impact measurement is measuring reach—how many individuals have experienced a piece and been impacted by a cultural output. This is done largely through large surveys and measuring footfall, internet traffic, ticket sales, social media analysis, and media mentions. This data is usually triangulated with the significance of impact to make a case for impact. However, this approach is problematic as reach does not necessarily equate to change. It can be a pathway to change but it isn't actually evidence of change. It's evidence that it's reached a number of people. Triangulating it with examples of significance from verbal testimonies, a report may make a more compelling argument, but as we saw, there are problems measuring significance; we can't really state this in any convincing way. The claim could be made that with higher numbers of engagement, there is a higher likelihood of change, but we're assuming this. We can't measure this in any truly meaningful way.

Adding to the unfeasibility and flawed nature of measuring changes on individuals, as touched upon before, those who work in the museum and university sectors tend to assume they have positive impacts. Even before impact measuring takes place, this is a fundamental flaw. Impact measuring is based on the assumption that research and cultural outputs, such as heritage projects and museums, inherently produce positive changes to individuals and society. This is compounded by the high stakes involved: final reports for heritage projects must specify how much positive impact was achieved, as funding has already been allocated. Similarly, research impact assessments argue for the significance and reach of this impact to secure further funding. In effect, what makes an impact on policymakers is if the change is countable, rather than if the nature of the change is actually positive or negative. This relates to a wider set of auditing processes designed to make research accountable to the public purse. This constitutes the 'implicit optimism' of the impact agenda. This pressure then distorts findings and methodologies from the outset, as impact can be both negative and positive. Alongside this external pressure, there is a largely shared belief within the museum and university sectors that they only produce good impact, driven by an almost crusading belief in the power of cultural productions to enlighten the masses. Rukšėnaitė (2022, p. 21) describes this as the "goodness regime," which relates to an unquestioned belief in the potent forces of culture and their power to fix societal evils such as sexism, racism, and other social ills. Belfiore and Bennett (2007, p. 135) refer to this as an "assumption" when discussing the social value of the arts and their performance impacts. They argue that the corollary of such thinking is that the arts produce no negative impacts, and if they do, they are not worth mentioning. These assumptions are just that assumptions and they have not been tested. As a result, methodologies that only measure and explore positive impacts are inherently flawed.

The Results, Where the Problem Lies

Two words: accountability and advocacy. The outline above hopefully serves as a stark reminder of the unfeasibility of accurately measuring changes in behaviour brought about by engagement with cultural outputs. None of these points are new; they have been extensively discussed in academic literature. Anyone with a research background would recognise the core issues surrounding the attempt to measure impact rigorously and meaningfully as described. So, why does this persistent obsession with measuring the impact of changes in behaviour in the culture sector continue?

My argument is that it's driven primarily by accountability and advocacy, manifested in the form of reports and assessments. Only by examining the final product—the report itself—can we truly understand the problem and consider effective solutions.

Before delving deeper, I want to invoke American philosopher Harry G. Frankfurt's definition of bullshit (Frankfurt, 2005) from his seminal book *On Bullshit*. Frankfurt describes two central aspects: first, 'mindlessness', a complete lack of concern with the truth on the part of the bullshitter; and second, the fact that behind any production of bullshit lies a bullshitter who is intentionally misleading his or her interlocutors so as to pursue his or her own interests and purposes.

This section focuses on this lack of concern for the truth, in this case, the flawed methodologies and results of measuring impact in the cultural sector to make claims around the alleged transformative powers of the arts and culture and their presumed positive impacts on behaviours.

The methodologies employed serve a performative purpose: advocacy and the appearance of accountability. They lend a veneer of legitimacy and seriousness, but— as we've seen—impact measurement from the start is fundamentally flawed. The focus on optics—on producing 'impressive' reports and workloads—tends towards what Frankfurt would call bullshit: superficial, misrepresentative, and ultimately disconnected from the truth.

My own experience is relevant here. In my role as an exhibition/heritage project evaluator, I was commissioned to evaluate whether projects had achieved their aims. These would be documented in a report, with the results often outlined in an executive summary using reporting language, sprinkled with some negative points that were largely couched in terms of "could do more of this"—not necessarily negatives. They have the feeling of a ritual, akin to a court offering, complete with presentation, graphs, and the accoutrements of data. But does anyone ever truly read them, though? They are not compelling; instead, they are largely vacuous reports, lent a serious air by graphs, charts, numbers, and formal sections with business language—such as the executive summary—given levity and attractiveness to busy cultural sector leaders and shakers through the use of colours, often overused (pink, teal, etc.), and the heart-warming quotes and photos of visitors or community groups from marginalised backgrounds, intended to lend the exercise a sense of wholesomeness.

It is truly an art—balancing seriousness, ego-stroking, engaging formats—and those who excel at it, such as leading evaluation consultancies, command high fees in the UK sector. But here's the paradox: do the readers of these reports—those complicit in the "bullshitisation"—believe what they see? Or are they lulled into a hypnotic state, mesmerised by buzzwords, self-promotion, and bland corporate language? Is this act of "bullshit" deliberate or simply an unconscious consequence of bureaucratic culture? In my view, the readers are often complicit. They may not believe in the content, but they participate in the ritual—accepting these reports as part of a performative dance that secures funding and maintains status. I sometimes wonder whether there is an increasing fetishisation of presentation. There's nothing inherently wrong with being attracted to graphs, presentations, and elaborate diagrams; different strokes for different folks, as Sly Stone once sang. But acknowledging this fetishisation—and understanding its role in maintaining the status quo—would be a step forward.

The real issue is that the actual data collection—the core process, often overlooked, involving fieldwork and data gathering, whether with or without a clipboard—is either avoided, poorly paid, or handed over to volunteers. The flawed nature of the methodologies and results is ignored. Instead, organisations focus on the spectacle of presentation—the visual and narrative packaging—or the planning phase, including setting objectives, targets, and elaborate theories of change. These plans are often incorporated into funding applications to demonstrate robustness and rigour, and so the elaborate performance continues.

Why does all this matter? Because truth matters. The waste of resources, the misallocation of people's skills, and the distortion of priorities—these exercises of bullshit are to the detriment of everyone involved, the sector, and ultimately all of us. Additionally, the obsession with impact measurement and its trappings, for the reasons outlined above, has serious effects on the cultural outputs themselves. This is the focus of the next section.

The Negative Effects of Impact Measurement on the Cultural Sector

This might be the most contentious and speculative part of my argument. It's about how impact culture affects creative outputs—exhibitions, heritage projects, research. From my own experience, and backed by thinkers like Belfiore and Fisher, impact culture favours the easily measurable, flattening the complexity of human expression. It's the interference of technocratic, managerial culture—turning arts and research into standardised, consumable products. Everything is becoming a "one show," a homogenised, inoffensive, flattened, commercially appealing spectacle. I believe impact measuring is part of the managerial, auditing culture that dominates, and it has played a major role in creating this environment. Personally, I would like to see more of the messy, half-formed ideas, the spontaneity, the humour, the anger, the whimsical, the directionless—in short, life and experimentation, rather than the self-promotion and polished, highly competitive environments that dominate today in the cultural sector. The successful cultural producer is increasingly rewarded for technocratic, managerial qualities, while those who can't or won't compete are sidelined.

I argue that the increasing instrumentalisation of the arts and research, which goes hand in hand with impact measuring, has led to greater division, as these fields are often employed to promote specific agendas rather than encourage meaningful dialogue. This can result in a backlash, with audiences reacting against overtly preachy narratives and simplistic portrayals in exhibitions. The assumption that every initiative has a positive impact overlooks the complexity of societal issues and contributes to public uncertainty, alienating the public who may feel their concerns are dismissed. This dynamic can stifle productive conversations and hinder collaboration.

Moreover, impact measurement tends to prioritise short-term, easily quantifiable results over deeper, transformative shifts, which ultimately sustains the hegemonic ideology by discouraging genuine political engagement and reinforcing dominant worldviews that are perceived as natural and unchangeable. Jeremy Gilbert and the late Mark Fisher have described the impact report in a university context as a ranking system—a control mechanism used by management. The report becomes a tool for bureaucrats, but there is a crucial distinction: effective management and good governance are not the same as managerialism. The latter—more akin to bureaucracy—entails a technocratic authority that often serves narrow interests, imposing regulatory processes that weaken organisational autonomy without genuinely enhancing efficiency (Gilbert & Fisher, 2013).

Additionally, this "bullshitisation" diminishes trust among visitors, researchers, and the public. It encourages self-promotion rather than meaningful knowledge, deep reflection, or societal change (Brown, 2018). In university research, impact assessments—especially under REF—become career-defining tools, turning inquiry into a game of compliance in favour of superficial metrics based on questionable data. The arts and humanities lose out in this impact-driven culture, which favours policy, utilitarian, and economic outcomes that can be more clearly evidenced in other fields. Impact assessments skew research priorities, marginalise experimental work, and reinforce a focus on industry and policy change.

Furthermore, measuring change within a project has its value to participants as a reflective exercise, but having it measured, quantified, and always massaged into a positive can distort authentic discoveries. An image control exercise can negatively impact trust building, which is crucial in community initiatives, and reinforces hierarchies. Communities often perceive their involvement as superficial, tokenistic, or instrumentalised—primarily for image-building. When the purpose shifts from the communities themselves to cater to funders or external interests, it can lead to detrimental short-termism. Reporting should ideally be a side product, primarily for advocacy. Moreover, impact evaluations are often dictated by national organisations, frequently disconnected from grassroots contexts. When evaluation is not rooted in grassroots experiences, it prioritises top-down metrics over genuine community engagement.

Finally, impact measurement, along with its complex jargon and infrastructure, creates additional barriers for community and marginalised groups to be included on an equal basis in cultural projects. It also consolidates power among those with the resources and language skills necessary to navigate and implement an impact agenda.

Potential Solutions, New Directions

There is nothing inherently wrong with cultural outputs that aim to have an impact. However, this chapter's argument is to recognise the limitations of measurement.

Advocacy and image control in the cultural sector have reached levels I have not seen in other European countries where I have lived. It has created a vibrant but heavily managed environment. Compare this to other European countries, such as Spain, where a quick glance at job advertisements reveals the difference. However, it is also clear that the prevailing over-managerialisation and auditing culture have negative effects. I believe it's time to learn from these mistakes and adopt a different approach—rather than simply copying the UK's advocacy-focused cultural sector.

Let me be clear, I don't suggest we stop creating or planning for impact. There are many positives, especially when based on rigorous research. However, I propose that we reflect honestly on the limitations of measurement and question the overreliance on the written word as the primary means of assessing impact. We should instead emphasise more research into how the public interacts with cultural outputs, how visitors interpret exhibitions and how the public engages with research. These can serve as valuable resources for the cultural producers themselves—a rich seam and a useful research source. Further research will identify a broader range of cultural framings used in relation to different exhibitions. Understanding these presuppositions and assumptions could, for example, help create exhibitions that draw from, challenge, or reflect on them, contributing to a growing body of case material—hopefully in insightful and provocative ways. That's where our efforts and resources should go if we truly want arts and culture to contribute to impact and active citizenship.

I view advocacy—whether for funding or presenting a case to the public—as a distinct process. This is where we gather data, providing clear causal links that will be limited in most cases—being honest about what can be proven—and limit our claims accordingly. We should be transparent about who it's for and what it aims to achieve.

Separately, assessments for funding and accountability should be based on the integrity and quality of the research or exhibition, judged by independent experts in those fields. Extra care should be taken to ensure the assessment process is as impartial and arms-length as possible, involving outside assessors from different sectors and rotating reviewers to minimise bias and favouritism. This process can be enriched by incorporating diverse perspectives, rather than becoming a rigid, flattening procedure. Respect for expertise is crucial.

I understand that publicly funded cultural outputs need accountability, but this should be tailored to each project. It must be flexible and not a one-size-fits-all approach. The process should not overshadow the actual output if the process becomes more valued or better funded than the work itself.

In many cases where measurement isn't possible, I'm calling for an end—or at least a limit, to impact measurement driven by short-termism, emphasising honesty about what can be measured, proven, and effectively presented. I believe we should admit it and refrain from using it. Honesty is key.

Glossary of Terms/Index

AHRC

Arts and Humanities Research Council—a funding body in the UK responsible for supporting research in the arts and humanities.

Cultural Sector in the UK

Includes museums, galleries, heritage sites, and arts organisations operating within the UK, often governed by national policies and funding structures.

Impact Assessment / Impact Evaluation

The systematic process of measuring and demonstrating the societal, cultural, or economic effects of a project or research activity, often tied to UK funding mandates.

Impact Case Study

A structured, typically three- to five-page report submitted by UK universities to demonstrate the societal, economic, or cultural impact of their research. Used in REF evaluations to assess research influence beyond academia.

Impact Officer / Impact Evaluator

A professional role within UK institutions responsible for designing, collecting, and reporting impact data to meet funding and accountability requirements.

Managerialism

A belief in or reliance on management techniques and principles to govern institutions, often emphasising efficiency and accountability.

National Lottery Heritage Fund

A UK government-administered funding body that distributes proceeds from the National Lottery to heritage, cultural, and community projects. Impact assessments are often required to demonstrate project success.

Research Excellence Framework (REF)

A periodic national assessment in the UK that evaluates the quality and impact of university research. Outcomes influence funding allocation and institutional reputation.

UKRI

UK Research and Innovation—the national funding agency that supports research and innovation across the UK.

References

Belfiore, E. (2009). On bullshit in cultural policy practice and research: Notes from the British case. *International Journal of Cultural Policy, 15*(3), 343–359.

Belfiore, E., & Bennett, O. (2007). Rethinking the social impacts of the arts. *International Journal of Cultural Policy, 13*(2), 135–151.

Bourdieu, P. (1984). *Distinction: A social critique of the judgement of taste* (R. Nice, Trans.). Harvard University Press.

Brown, W. (2018). *Notes on creative practice research in the age of neoliberal hopelessness*. University of Bedfordshire.

Fisher, M. (2009). *Capitalist realism: Is there no alternative?* Zero Books.

Frankfurt, H. G. (2005). *On bullshit*. Princeton University Press.

Gilbert, J., & Fisher, M. (2013). Capitalist realism and neoliberal hegemony: A dialogue. *New Formations, 80*(1). https://doi.org/10.3898/neWF.80/81.05.2013

Graeber, D. (2018). *Bullshit jobs: A theory*. Simon & Schuster.

Hooper-Greenhill, E. (2000). *Museums and the interpretation of visual culture.* Routledge.

Hooper-Greenhill, E. (2006). Studying visitors. In S. Macdonald (Ed.), *A companion to museum studies* (pp. 123–136). Blackwell.

Macdonald S (2002). *Behind the scenes at the Science Museum.* Berg Publishers, Oxford.

Rukšėnaitė, G. (2022). *Balancing between artistic freedom and instrumentalisation: Art's role in Nordic cultural policy* [Master's thesis, Aalto University]. AaltoDoc. https://aaltodoc.aalto.fi/handle/123456789/115598

CHAPTER 4

Exploring Cultural Impact in Exhibition Design: A Holistic Approach to Sustainability Across the Life Cycle Phases

Laura Asión-Suñer and Anna Maria Biedermann

The Role of Culture in Sustainability Development

Incorporating the cultural dimension into the traditional three pillars of sustainability—environmental, economic and social—is essential for achieving a truly holistic approach to sustainable development (Soini & Birkeland, 2014). Culture is a cornerstone of identity and human development, shaping values, behaviors and decision-making processes (British

L. Asión-Suñer (✉)
Department of Design and Manufacturing Engineering, University of Zaragoza, Zaragoza, Spain
e-mail: lauraasion@unizar.es

A. M. Biedermann
Department of Design and Manufacturing Engineering, University of Zaragoza, Zaragoza, Spain

OAAEP Group, Funded by the Government of Aragón with FEDER Funds, Zaragoza, Spain
e-mail: anna@unizar.es

A. M. Biedermann, F. J. Galán Pérez (eds.), *Interconnections between Culture and Sustainability*,
https://doi.org/10.1007/978-3-032-21061-6_4

Council, 2020). Cultural projects influence both territorial and social development (Jelinčić, 2022). By integrating cultural considerations, sustainability initiatives can be tailored to the specific contexts and traditions of the communities they serve, ensuring relevance and inclusivity. Cultural practices often carry deep ecological knowledge and traditional wisdom that enhance resilience to environmental challenges, while fostering innovation and creativity through unique expressions and approaches (George, 2024). It is also the culture that shapes our views toward sustainability (Birtch, 2024).

Additionally, culture plays a crucial role in promoting inclusivity and social cohesion, bridging divides in multicultural societies and ensuring that sustainability efforts address representation and equity (Pascual, 2016). Recognizing culture also supports intergenerational equity by preserving and passing down traditions, knowledge and identity, which are fundamental to long-term sustainability (Throsby, 2017). Furthermore, cultural expressions contribute significantly to well-being by providing purpose, belonging and joy, reminding us that sustainability is not just about survival but about well-being and thriving. Aligning with global agendas such as the UNESCO Culture 2030 Indicators and the Sustainable Development Goals (UNESCO, 2019), particularly SDG Target 11.4, which emphasizes safeguarding cultural heritage, integrating culture enriches sustainability efforts by honoring diversity and addressing the unique needs of communities. By linking culture with environmental, economic and social dimensions, we can create sustainable solutions that are not only effective but also meaningful and enduring (Biedermann et al., 2022).

When discussing the inclusion of the cultural dimension in sustainability within the literature, many studies focus on generating actions aimed at making services or products more sustainable without clearly providing tools to measure how much more sustainable a product or service becomes as a result of those recommendations (Soini & Dessein, 2016; Hawkes, 2001; Brocchi, 2008). While known strategies from sustainable design methodologies can be applied, we cannot confidently assert that a product or service is indeed more sustainable until it is evaluated in a specific context. For instance, consider replacing printed materials with projections, which, in theory, aligns with dematerialization and reducing the use of raw materials. However, depending on the duration of the event, this substitution could ultimately be less sustainable—generating a greater environmental, economic and/or social impact—than using traditional printed materials (Muñoz et al., 2021).

The Challenges of the Sustainability Assessment in Four Dimensions

To integrate the four dimensions of sustainability (environmental, social, economic, and cultural), an assessment of the impacts generated by each of them is required. Currently, there are frameworks that address this assessment, such as the *Impact & Insight Toolkit* (Booth-Kurpnieks et al., 2023), developed in England, which is based on the Australian tool TAKSO developed by the Cultural Development Network (CDN, 2025). Both models present an approach based on the four dimensions of sustainability mentioned and an additional fifth dimension: governance. This represents a breakthrough in the development of sustainability assessment methodologies.

However, both tools have two important limitations. The first is that they rely on declarative perception as a measure of impact. Although this approach is appropriate for measuring cultural impact, it has limitations in the other three dimensions. If we focus on the environmental dimension, these tools assess whether public understanding of ecological problems has improved using Likert scales. Yet, environmental assessment conventionally employs standardized quantitative indicators, such as kgCO_2 equivalent for Global Warming Potential or megajoules (MJ) for embodied energy. Similarly, in the economic dimension, the toolkit evaluates the perception of improvements in professional practice capability, whereas established economic indicators include metrics such as total costs (expressed in monetary units) or added value.

The second limitation lies in the toolkit's linear conceptualization of cultural projects. Its implementation follows a process based on predefined Key Performance Indicators (KPIs) for planning cultural actions. These KPIs are implemented and subsequently evaluated through surveys. Therefore, this process does not take into account the incorporation of a life-cycle perspective, which is essential for conducting a holistic assessment. These limitations demonstrate the need for further development of sustainability assessment methodologies in the cultural field.

Based on these four dimensions, every product, action, or flow involved in the life cycle of the analyzed system (depending on the scope) should ideally be assessed simultaneously across all four dimensions using the indicators that give the major understanding about the impact produced. This comprehensive perspective ensures that sustainability is addressed not only in terms of ecological preservation, financial viability, and social equity, but also through the recognition and safeguarding of cultural values. In this regard, the cultural dimension poses a significant challenge when approaching a quantitative assessment in a similar way to how the other three dimensions do in a standardized way (Biedermann et al., 2024).

Thus, integrating cultural considerations into sustainability assessment requires a flexible and participatory approach capable of measuring quantitative indicators as well as subjective values (Stephenson, 2023). In the literature, a wide range of proposals for evaluating cultural impact is found. Some methodologies are designed to document and showcase efforts made in cultural activities (O'Brien & Oakley, 2015), while others aim to compare the initial situation with the improved service (Biedermann et al., 2023). These approaches are useful for measuring the effectiveness of individual projects or providing accountability for cultural initiatives.

However, what is notably absent is a framework capable of maintaining a certain level of objectivity while enabling the comparison of different cultural products or services across multiple dimensions. Existing methods often focus on qualitative insights or context-specific evaluations, which, while valuable, are not easily adaptable to different scenarios or comparable across diverse cultural contexts (Sagnia, 2004). The challenge lies in developing a standardized yet flexible approach that incorporates the four dimensions of sustainability—environmental, economic, social and cultural—while ensuring that the cultural dimension is given equal weight and treated with the rigor typically afforded to the others, and analyzed in the life cycle perspective. This would require methodologies that go beyond anecdotal or context-specific measurements, focusing instead on creating scalable, transferable indicators that balance quantitative and qualitative data.

Cultural Sustainability Assessment Indicators

The indicators used in some studies to define cultural impact often, by their very nature, reference another dimension (Blumenthal & Wittlin, 1971; Jelinčić, 2021; Yan & Liu, 2023). For example, quantitative indicators include metrics such as the number of participants or hours of cultural training delivered (more representative for the social dimension) or funds allocated to cultural preservation (proper to the economic dimension). This overlap is not uncommon, as we also observe similar crossovers in the evaluation of other dimensions, for instance, salaries used as indicators of social impact. We are not suggesting that this is inherently a mistake, as the four dimensions are deeply interconnected, and it is natural for this interconnection and interdependence to be reflected in the nature of the indicators themselves. However, the issue arises when none of the indicators employed to define a specific dimension is genuinely capable of assessing the impact rooted in that dimension. This lack of precision undermines the ability to accurately evaluate the distinct contributions of cultural impact.

The evaluation process is often laborious and time-consuming; therefore, it is common to limit the number of indicators within each dimension. This raises the question of whether it is appropriate to narrow the scope within the cultural dimension—particularly when it is already expressed using a Likert scale—or whether it would be more meaningful to construct a composite indicator. The purpose of the results is not only to understand the impacts—that is, the changes produced—but also to enable comparison across different services and to support improvement by introducing adjustments that reduce negative impacts or enhance positive ones.

Each cultural service operates under specific needs and within a distinct contextual environment. Actions and strategies that prove beneficial for one may be detrimental for another when variables such as the space, durability, or the nature of the stakeholders involved are considered. Applying a standardized methodology is essential for objectively comparing the impact of different cultural services. This sustainability assessment should evaluate the same indicators at the same life cycle phases, thus involving the same stakeholders. These parameters facilitate the comparison of results, especially if the same units of measurement are used to conduct these assessments.

Figure 4.1 compares three different methods for cultural sustainability assessment:

1. *Method based on Life Cycle Sustainability Assessment (LCSA), which includes Cultural Dimension* (Biedermann et al., 2024);
2. *Impact & Insight Toolkit* (Booth-Kurpnieks et al., 2023);
3. *TAKSO tool* developed by the Cultural Development Network (CDN, 2025).

For these three cases, the indicators employed by each method have been compiled with the aim of identifying common elements that enable establishment of comparison. These indicators encompass both qualitative and quantitative dimensions and can be assessed through a range of

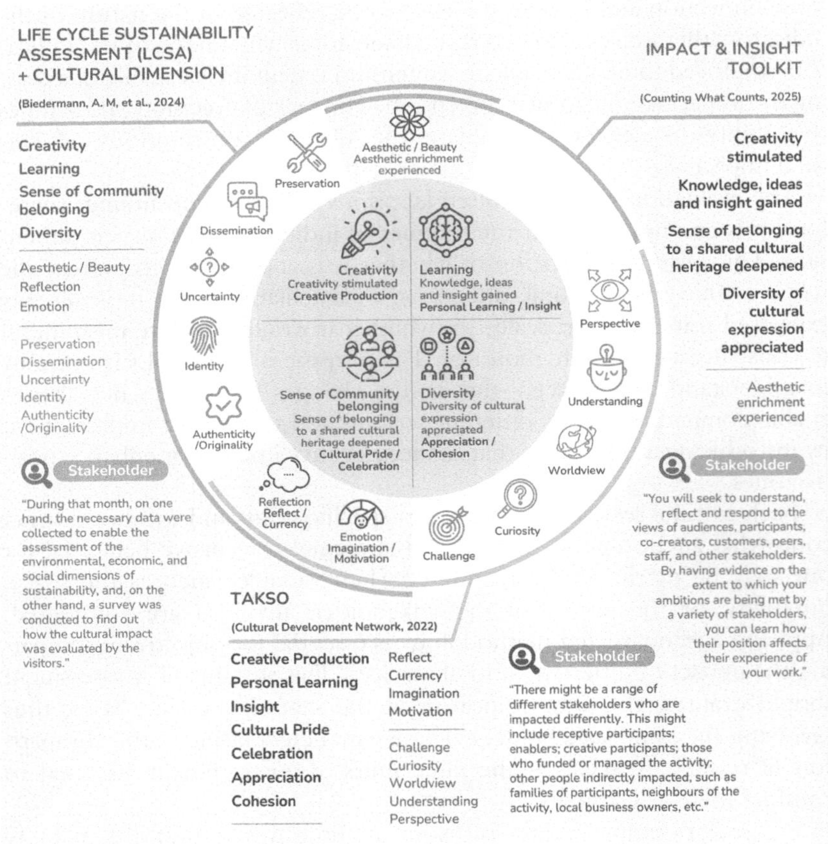

Fig. 4.1 Comparison of three methods for Cultural Sustainability Assessment

methodologies, including interviews, participatory workshops, and surveys, among others. Additionally, each method specifies the stakeholder groups it targets, establishing a direct relationship with the life cycle phases of the cultural service, as each stakeholder is typically associated with a specific stage within that cycle.

Four common indicators have been identified across the three assessment methods: creativity, learning, sense of community belonging, and diversity. In the case of the TAKSO method, learning is divided into two distinct indicators—personal learning and insights—while sense of community belonging is similarly represented by both cultural pride and celebration. Although the method introduces subtle distinctions between these indicators, both pairs ultimately correspond to the common indicators shared with the other two assessment frameworks.

If the objective is to evaluate a single representative indicator for the cultural dimension, it is therefore advisable to select one of these four indicators common to all three methods. Doing so would help to ensure comparability of results, provided that the assessments use the same units of measurement and engage equivalent stakeholder groups.

Additionally, certain indicators are shared between pairs of methods. For example, reflection and emotion are common to both LCSA + Culture and TAKSO, while aesthetics appears in both LCSA + Culture and Impact & Insight. Notably, all five indicators included in the Impact & Insight method correspond with those in the other two frameworks. In contrast, both LCSA + Culture and TAKSO define several unique indicators not found in the others. Specifically, LCSA + Culture incorporates preservation, dissemination, uncertainty, ***identity***, and ***authenticity/originality***, while TAKSO introduces ***perspective***, ***understanding***, ***worldview***, ***curiosity***, and ***challenge***.

Additionally, Fig. 4.1 presents selected direct statements from each method regarding the stakeholders they consider. Although both TAKSO and Impact & Insight conceptually address a general public, in practice their surveys and participatory workshops focus primarily on the visitor audience. The same applies to LCSA + Culture. While this method does not restrict the range of stakeholders, in the case study analyzed in Fig. 4.1, it is likewise applied exclusively to visitors. As a result, all methods ultimately concentrate their evaluations on the same stakeholder group within the provision phase, despite the theoretical potential to incorporate additional stakeholders from across the life cycle. This outcome is unsurprising, given that visitors are typically regarded as the stakeholder group upon whom cultural services exert the most direct and observable cultural impact. Nevertheless, cultural services also influence other stakeholders throughout the life cycle, including co-creators, technical staff, and operational teams. For this reason, it is necessary to advance cultural impact assessment methodologies to ensure that indicators are evaluated not only in relation to visitors but also to other stakeholders involved in and affected by cultural services across all life cycle phases.

Life Cycle Sustainability Assessment Perspective

Current studies have successfully assessed the cultural dimension through quantitative data obtained during the use phase through visitor surveys. The results are integrated with the assessment of sustainability in the environmental, economic, and social dimensions during the cultural service provision phase (Biedermann et al., 2024). However, there are no previous cases of a holistic assessment of the four dimensions that consider the entire life cycle. This significant gap highlights both the complexity and the pressing need for innovative, context-sensitive approaches to assess cultural impacts beyond the use phase, particularly in the creation and end-of-life phases, which have remained largely unexplored in sustainability research. In particular, the creation phase holds substantial cultural value, as it involves processes of curatorial negotiation, design decisions and the activation of local and institutional knowledge systems that shape the cultural significance of a cultural product (Fairclough et al., 2008). Yet, these formative processes are seldom subjected to systematic evaluation, despite their capacity to influence cultural narratives, representation and inclusivity long before audiences interact with the final product (O'Brien & Oakley, 2015).

For this reason, this chapter proposes a paradigm shift, emphasizing critical, though often overlooked, cultural impacts that are generated and experienced during the creation, provision, and end-of-life phase of a cultural service. By addressing this critical yet underexplored perspective, the study aims to contribute to the development of methodological approaches and context-sensitive indicators that facilitate more integrated and balanced assessment of sustainability across its four dimensions along the entire life cycle.

Creation Stage

This initial stage is not merely preparatory, but rather a dynamic site of cultural production and transformation involving artists, collaborators, and sometimes early community participants. Cultural events, conceived to "celebrate and promote artistic, historical, and social aspects," begin to generate their impact from the moment of conception and initial coordination, which implies that impact considerations are essential from the outset.

The impacts that emerge during the creative process are of a unique nature. They can manifest in various ways—for example, in the creators themselves (skills development, shifts in perspective, new understandings), in collaborative dynamics (formation of new social bonds, shared languages, trust), and in the incipient cultural forms being developed (innovative artistic expressions, reinterpretations of heritage). These "processual impacts" are often intangible and relational, and they shape both the final event intended for the public and its potential for broader cultural resonance. Culture-based creativity—which involves intuition, affect, and the challenging of conventions—can lead to new visions and community values, first manifesting among the creators (KEA European Affairs, 2009).

Artists, far from being isolated geniuses, are "embedded in social and cultural contexts," and their work is "influenced by the social and historical conditions in which they live." Interactions, social norms, and institutional contexts within the creative process shape what is considered "creative" and "impactful" (Burns et al., 2015). Even before public presentation, the act of creating has the capacity to challenge norms and generate cultural identities (Tchermalykh, 2024). In this way, the creative process immerses those who participate in it in an experience that fosters their intellectual, social, and emotional growth. Its evaluation must therefore consider the observation of personal motivation and commitment.

This approach is inspired by thinkers such as Douglas Worts, whose work advocates for a vision of culture as a "living process" of human adaptation and meaning-making (Worts, 2011). Evaluating the cultural impact in the creation phase requires understanding how the cultural event is configured from the beginning, and can become "catalysts for cultural adaptation and transformation" (Worts, 2004). Therefore, carrying out a prospective assessment involves anticipating potential impacts, while formative evaluation allows for iterative refinement based on emerging insights during the creative process. Worts' "Inside-Outside Model" (Worts & Dal Santo, 2022) encourages planning for cultural impacts from the beginning, making it directly relevant for evaluation in the creation phase. Its emphasis on "meaningful experiences" suggests that the creative process should be a site of experiences for those involved (Worts, 2003).

The collaboration and knowledge transfer carried out during the creation phase allows for the generation of a cultural experience that includes a broad audience. Thus, cultural events seek to generate dialogue between different stakeholders, with the creation phase being the one with the greatest interaction and meaning-making within a core group (Burns et al., 2015). The values and dynamics represented within this core group—whether collaborative, hierarchical, or inclusive—will inevitably shape the event's "DNA." Therefore, if the creation phase itself embodies positive cultural values (e.g., respect, shared voice), it is more likely that the final event will successfully project these values and achieve its intended cultural impact (Impact & Insight Toolkit). This makes the creation phase a kind of rehearsal or cultural prefiguration. Dimensions such as fostering "appreciation and engagement with new and diverse cultural experiences, expanding knowledge and understanding, and stimulating creativity" (Wallace, 2024) may first emerge within the creative team itself. The framework of "shifts in meaning, shifts in connection, and shifts in power" (Vervoort et al., 2024) is highly relevant for conceptualizing these early impacts.

Furthermore, assessing cultural impact in the creative phase can alter the creative process itself. This is known as the "observer effect" and poses a particular challenge, as the assessment introduces an element of scrutiny (Zhang, 2024). During evaluation, creators alter their behavior and decisions according to their expectations of what evaluators want to see. Whether consciously or unconsciously, this alteration can affect creativity and the interpretation of impacts. For this reason, evaluation processes should be highly participatory and responsive. This helps build trust and ensure that the evaluation supports the creative process.

The quantification and evaluation of cultural impacts during the creation phase present distinct methodological and conceptual challenges. This is because some participants not only register cultural impacts, such as acquiring new knowledge, shifting values, or developing a sense of belonging, but also actively shape the elements that will, in turn, influence the cultural experiences of others. Nevertheless, cultural impact can occur in every participant in this process; therefore, if we aim for a holistic evaluation, we must take it into account.

Provision Stage

This stage refers to the moment when the public engages with the cultural service or enjoys a cultural product. In this context, audiences are the primary stakeholders. In numerous previous studies, it is precisely the audiences who, after attending the visit or cultural event, respond to a series of questions, assessing the degree of impact that the cultural service has had on them in aspects related to them as individuals, or part of a collective or in relation to heritage etc.

If we are interested in conducting a more segmented study, we could divide our audience according to their characteristics by applying various methods. For instance, Falk and Lynn (2008) and Falk and Dierking (2016) proposes identifying different types of visitors based on the needs they seek to fulfill—that is, their underlying motivations. Visits are not random but are driven by a purpose, even if that purpose is not always consciously recognized. Falk primarily distinguishes five visitor types: Explorers, Facilitators, Professional/Hobbyists, Experience Seekers, and Rechargers/Reflective visitors.

Other commonly applied segmentation criteria are based entirely on demographic characteristics such as geographic origin, age, or educational level. Further dimensions include frequency of visits or the composition of the group during the experience. By analyzing cultural impact through a segmented audience approach, we can draw detailed conclusions about its effects on specific groups while also gathering data on overall sustainability.

However, along the provision stage, it is also essential to consider all the personnel involved in visitor service before, during, and after their experience, such as ticketing staff, front-of-house personnel, or staff working in gift shops. Internal security personnel are also continuously exposed to the cultural content for extended periods. Likewise, cleaning and maintenance personnel—responsible for safeguarding the integrity of collections and exhibition layouts—can also be expected to establish a form of relationship with them through ongoing interaction. Assuming that these occupational groups are unaffected by cultural impact would be a significant oversimplification. Cultural impact can emerge not only through direct exposure to content but also through the way work is structured and experienced in this phase.

Additionally, the marketing team plays a key role at this stage by implementing strategies planned earlier and maintaining constant communication with the media, which in turn shapes public perception of the cultural value of the service provided.

End-of-Life Stage

The end-of-life stage is not merely a terminal event, but a moment of profound transition that can reconfigure collective memory, values, and the identity of a community (Lowenthal, 1985). While the circular economy has popularized strategies such as recycling, reuse, and remanufacturing for material aspects (Ellen MacArthur Foundation, 2013), a truly sustainable approach must also consider what happens to the inherent cultural value of what is dismantled, discarded, or ceases to exist in its original form. This cultural value is not always tangible; it may reside in symbolic meaning, associated narratives, the social practices it supported, or the skills it helped to foster. It is at this stage, when a cultural service or product reaches its end, that decisions must be made about preserving elements of tangible or intangible heritage, archiving them, and managing their future. In many cases, technical teams are also involved in dismantling the infrastructures that supported the cultural service or product. They must determine whether components can be reused or should be discarded. These professional teams, to varying degrees, come into contact with the cultural product or service. Even if they are not directly exposed to its content or did not participate in the cultural experience, this interaction can nevertheless spark their curiosity or motivation to engage with future cultural events.

Cultural sustainability in the end-of-life phase compels us to look beyond the material and the immediate. This makes it possible to consider the durability of the knowledge transmitted by the cultural product or service, valuing the memorability and intangible legacy of cultural creations. Integrating cultural impact assessment in this phase entails acquiring a new perspective where the end of the cycle is not a total loss, but a culturally enriched transformation. This is an opportunity to build new cultural services that are conscious and respectful of heritage.

Considering the end-of-life phase therefore involves actions that lay the groundwork for restarting the cycle, resulting in a repetitive process. Common events such as travelling exhibitions, concerts or theatre performances have a cyclical nature where the end-of-life phase is only partial. In this way, incomplete end-of-life stages relate to the creation phases of subsequent cycles, the scope of which may be more limited compared to the initial creation.

When discussing end-of-life, we must also confront the question of what criteria define the boundaries of preservation. As a society, can we truly preserve everything we consider heritage for an infinite amount of time? This question, perhaps, belongs in another book.

Case Analysis: Phases of Itinerary Exhibition Life Cycle and Stakeholders Involved

In order to translate these reflections into practical terms, they will be illustrated through a case study: a traveling exhibition as a cultural service. This example has been selected due to the complexity of the service, which involves the interaction of multiple stakeholders across varied geographical and cultural contexts, as well as distinct operational demands at each venue.

To carry out a life cycle sustainability assessment, it is essential to compile a comprehensive inventory that includes all actions, as well as the raw materials and equipment used in each phase of the service. Each of the elements—whether actions or materials— can generate environmental, economic, and social impacts, and, as we will demonstrate below, also cultural impacts.

If the life cycle of the travelling exhibition is divided into the three general phases (creation, provision and end-of-life), we observe that each of them involves distinct processes and stakeholders. Their actions generate concrete impacts and can be evaluated from a sustainability perspective. In this way, the stakeholders of a cultural service can not only generate cultural impact, but also come to receive it. Although visitors are traditionally considered as the main stakeholders, it is noted that other stakeholders such as museographers, designers or installers are also equally exposed to the cultural impact of the exhibition.

Figure 4.2 presents the structure and stakeholders involved in the general life cycle phases that comprise a cultural exhibition.

The **creation phase** plays a decisive role in shaping the cultural significance of an exhibition, as it is during this stage that key conceptual, curatorial, and logistical decisions are made. The creation phase encompasses the processes from the initial conception of the exhibition to the completion of its physical assembly. It is divided into four sub-phases:

1. *Planification*, led by organizers and administrators, who are responsible for managing timelines, resource allocation, partnerships, and compliance with legal or institutional frameworks. Their role is

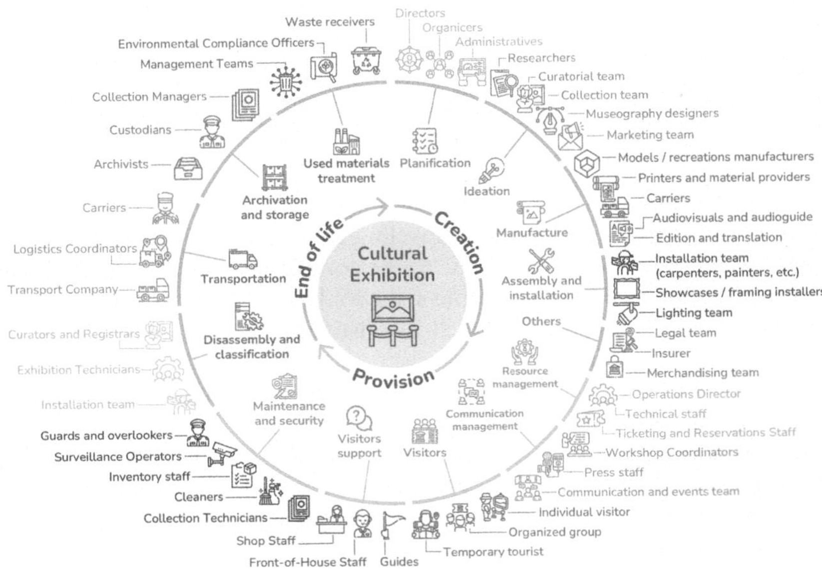

Fig. 4.2 Life cycle of a cultural exhibition: Phases and stakeholder involved

crucial, as they determine the composition of the working teams and have the capacity to promote diversity, interdisciplinarity, and balance within the project structure.

2. *Ideation* involves a diverse range of stakeholders, including artists, directors, curators, and commissioners, who are responsible for defining the exhibition's themes, selecting the narratives to be presented, and shaping the cultural messages to be conveyed and considering how the narrative might resonate across different future local contexts. At the same time, designers and museographers must interpret these contents in order to distribute them spatially and represent them through tangible elements, considering at all times the production process. This is a key task in order to get the content to the public in an effective way. In addition, they also consider the rotation of the exhibition and its next destination, taking into account factors such as transport. This directly influences the subsequent environmental and economic impacts.

3. *Manufacture* is carried out by specialized providers responsible for producing and transporting the exhibition components, including artworks, scenography structures, and informational supports. While this subphase is fully implemented at the first venue, it must be partially replicated at each subsequent destination. The possibility of designing museography elements to be easily transported and reused has a direct impact on all dimensions of sustainability.
4. *Assembly and installation*, which involves carpenters, installers and painters tasked with physically setting up the exhibition space according to the approved design.

The stakeholders involved in these four sub-phases are also supported by legal and insurance services, as well as by the teams responsible for merchandising. Defining "cultural impact" in the creation phase involves focusing on the changes that occur internally within the creative process and among its direct participants.

The **provision phase** covers the period during which the exhibition is open to the public. It includes:

1. *Resource management*, carried out by operations managers and their technical teams, as well as by the team responsible for ticketing and group reservations. Coordination is also required with the educational teams that deliver activities and workshops related to the exhibition content.
2. *Communication management*, which encompasses advertising efforts to promote the exhibition, liaising with press and media to ensure visibility and sustained interest in the subject matter, as well as organizing events aimed at maintaining public engagement throughout the entire period the exhibition remains open to visitors;
3. *Visitors*, who are the main audience to whom all the efforts invested during the creation phase of the exhibition are directed. For this reason, this group receives a high cultural impact no matter how the visit has taken place, whether as individuals or groups, or as close members of the community or spontaneous visitors. This makes visitors active recipients who contribute to the cultural value of the exhibition through their interpretations and interactions, influencing their future interactions and those of other potential visitors.

4. *Visitor support*, involving guides, informers and user experience personnel who mediate between the exhibition content and the public, thereby influencing audience perception and cultural impact. These professionals may also facilitate or adapt access to the exhibition content to accommodate the individual needs of visitors. The team responsible for the exhibition shop ensures access to objects and memorabilia that serve as tangible extensions of the experience, whether as souvenirs or as resources that allow visitors to deepen their understanding beyond the exhibition itself, such as through publications available for purchase.
5. *Maintenance and security*, which encompass activities carried out by collection managers and conservators, as well as cleaning and inventory staff. Together with security staff, these stakeholders ensure that the condition of the exhibition, its collections, museography elements, and the physical space in which it is presented are preserved and maintained in optimal condition throughout the entire provision phase.

The **end-of-life** phase addresses the processes that follow the exhibition's closure:

1. *Disassembly and classification*. Before the exhibition ends, it is essential to document the cultural impact that the event has had on visitors throughout the exhibition. This involves the collection of oral testimonies, the recording of designs or the development of user surveys, among other types of relevant documentary record. The final objective is to generate a cultural memory that not only considers the material aspects of the exhibition, but also intangible aspects such as its meaning and the way it is interpreted by different communities. This curatorial and archival work unfolds in parallel with the physical disassembly of the exhibition, carried out by carpenters, technicians, and logistics teams, once the collection has been safely removed from the exhibition space. Together, these efforts ensure that the conclusion of the exhibition is not merely a logistical task, but a culturally meaningful phase within its life cycle.

2. *Transportation* follows the disassembly. Couriers oversee the return of collections and artworks to their institutions of origin or the next venues on the exhibition's itinerary. Other components are transported to storage facilities for potential reuse, while materials designated for disposal are sent to appropriate recycling centers.
3. *Archivation and storage*, the management of archives, and the storage of collections require ongoing work after the exhibition ends. In this phase, the condition of the artworks is assessed, and, when necessary, restoration processes may be initiated. These tasks are overseen by collectors, conservators, and custodians, who ensure the long-term preservation and care of the items.
4. *Treatment of used materials*, handled by waste receivers and managers.

Though often overlooked in sustainability assessments, this phase generates significant cultural, environmental, economic, and social impacts, particularly in terms of heritage conservation, memory transmission and the sustainable management of resources.

A Holistic Approach to Sustainability Assessment

To comprehensively evaluate an itinerant exhibition service and from a life-cycle perspective, each action and material used should be inventoried for subsequent assessment across the four dimensions of sustainability. Naturally, some data will be difficult to obtain due to their nature. For instance, in traditional life cycle sustainability assessments, measuring the social impact of raw materials is notoriously complex. Similarly, evaluating the cultural impact of the raw material presents comparable challenges, not because it is impossible, but because of a lack of data or difficult access to data collection.

However, systematically recording and later evaluating the majority of actions carried out by stakeholders, as well as the materials used throughout the various phases of the life cycle, remains the only viable way to ensure a holistic evaluation. This approach allows for a more detailed understanding of the impact of cultural products and services on sustainability. Thus, by identifying the sources of impact, strategies can be proposed to mitigate the negative effects or enhance the positive ones.

Table 4.1 presents a framework for the holistic assessment of sustainability based on the Life Cycle Sustainability Assessment (LCSA) methodology. Originally developed to evaluate the environmental dimension (E-LCA), this approach was later expanded to incorporate economic (LCC) and social (S-LCA) dimensions. Today, LCSA is widely applied across multiple fields by both researchers and practitioners as a

Table 4.1 LCSA methodological scheme applied to the case of traveling exhibition

LCSA + Culture	*Description*	*Traveling exhibition case*
1. GOAL AND SCOPE DEFINITION	– Study objectives – Scope of the study – Functional unit definition	Objective: To identify the sources of impact of the traveling exhibition. Scope: Eight years of duration across eight different venues, covering the entire life cycle. Impact unit: Impact per year per visitor.
2. INVENTORY ANALYSIS	Comprehensive record of actions, materials, and energy, as well as the stakeholders involved.	Recording resources and actions from the start of planning throughout the entire life cycle, including transportation between the different venues, and considering resource management during the final phase.
3. IMPACT ASSESSEMNT	– Selection of indicators – Assessment tools – Evaluation of the impact of all actions or resources in the four dimensions	– Environment – kg CO_2 – Economy – € (euros) – Social – h (hours of work or involvement) – Culture – Likert scale Integration of quantitative, semi-quantitative, and qualitative data. Assessment of each action or resource across the four dimensions.
4. RESULTS INTERPRETATION	The interpretation of the results by competition with another product or services or to improve the service in the future.	Analysis of positive and negative impacts in each phase of the life cycle. Comparison of results with other cultural services. Improvement proposals based on the reduction of negative impacts and the enhancement of positive impacts.

standardized tool for assessing the sustainability of various systems and products. Owing to its broad acceptance and established methodological consistency, LCSA provides an appropriate platform for integrating culture as a fourth pillar of sustainability. In the proposed framework, the cultural dimension is introduced as an additional dimension alongside the existing three. Figure 4.3 illustrates the interrelationship between these four dimensions, the diverse range of potential indicators associated with each, and possible representative indicators selected for every dimension,

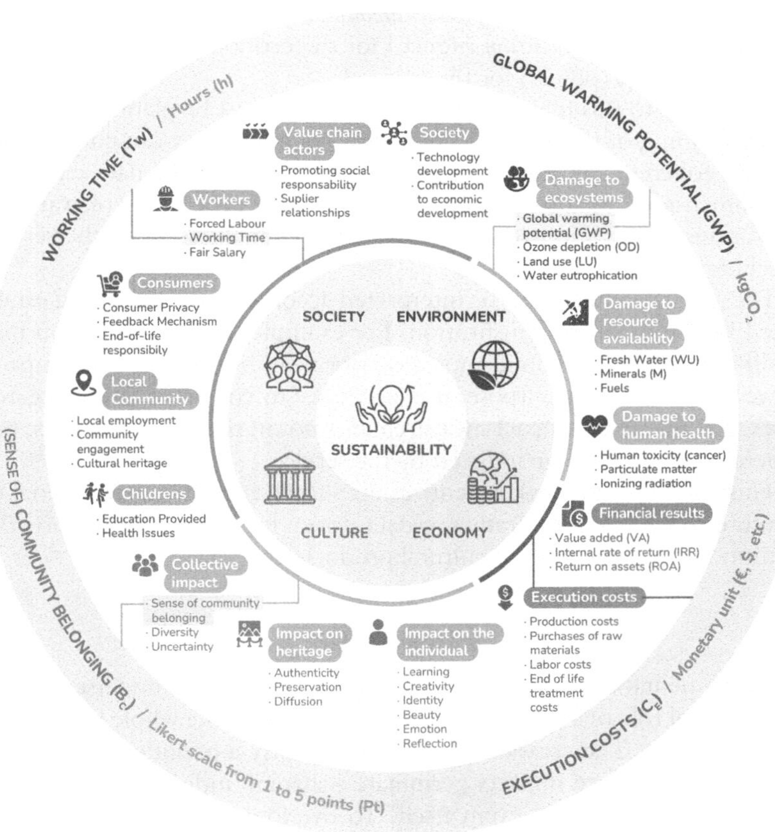

Fig. 4.3 Four dimensions of sustainability, with examples of possible indicators and units

together with the units of measurement in which the results may be expressed.

LCSA methodology is fundamentally based on quantitative data, which facilitates direct and comparable assessments across different dimensions. However, it also allows for the integration of semi-quantitative and qualitative methods where necessary. This characteristic promotes the effective integration of multiple dimensions within a single evaluation framework, by ensuring the consistent use of standardized units of measurement appropriate to each dimension. As illustrated in Fig. 4.3, the proposed case study applies one specific indicator to represent each dimension: *working time* (hours) for the social; *global warming potential* ($kgCO_2$) for the environmental; *execution costs* (€) for the economic; and *sense of community belonging* (points) for the cultural.

According to this framework, the evaluation must be planned from the very beginning of the service's creation; only in this way can comprehensive and coherent data collection be ensured. By measuring data across the four dimensions, we acknowledge throughout the process the relevance of the various types of impact to be assessed and analyzed once the service concludes.

The results obtained can be interpreted according to a weighting established by the organizing institution. For example, the interpretation may be different if the economic impact is prioritized over the cultural impact, or vice versa. For this purpose, it is necessary to construct an aggregated index that integrates impact measurements not in their original units, but as percentages of change induced by the service.

Thus, sustainability assessment results should not be viewed as final or absolute conclusions, but rather as data-driven feedback that supports the continuous improvement of cultural products and services.

Conclusion

The argumentation developed throughout this chapter underscores the fundamental role of assessing cultural impact during the whole life cycle of cultural events. The creation phase is not merely a prelude, but rather a fertile ground where impacts germinate—shaping individuals, collaborations, and the emerging event itself. To overlook this phase is to forgo crucial opportunities for refining, and amplifying cultural value. Equally important is the consideration of the end-of-life phase, which involves critical decisions regarding the archiving, dismantling, and potential reuse of cultural resources.

A comprehensive cultural sustainability assessment must address not only the immediate, visible impacts during provision but also those that emerge before and persist after the audience's experience. Integrating creation, provision, and end-of-life phases within a holistic, life-cycle framework offers a more rigorous evaluation of the cultural value. This approach requires methodologies that capture qualitative aspects of the creative process. By recollecting specific cultural indicators, involving stakeholders, and integrating these elements into existing sustainability assessment frameworks, it is possible to evaluate cultural impact more effectively.

Existing cultural impact assessment frameworks often focus on the provision phase, where audience reception provides observable data. The creation and end-of-life phases lack direct interaction with the audience, making it difficult to assess the cultural value generated during this period. Nevertheless, it involves activities such as ideation, production, archiving and dismantling. During the service provision phase not only, the audience should provide the cultural impact but the workers and collaborators as well. Their impact might differ, but they provide valuable information about cultural experience. If we pretend to strengthen culture that enables a diversity of points of view, the assessment process should also share this characteristic.

When examining the LCSA methodology, it becomes evident that certain indicators and dimensions, such as the number of working hours in the social dimension, already incorporate measurable data within the creation and end-of-life phases. If this is standard practice for other dimensions, why should it not be equally possible for the cultural dimension? To achieve a truly holistic and integrated approach, it is essential to consider all phases of the life cycle, rather than limit evaluations to the visitor experience during the provision phase.

Participatory methods reframe evaluation from an external judgement to an integral part of the co-creation of value. This carries ethical implications, moving toward a more equitable relationship between evaluators, creators, and communities. Shared evaluation and iterative feedback mean that evaluation not only measures value, but actively contributes to its development. This transforms the role of evaluation from the obligated reports that "judges" the cultural activity into one of support and development, collaborative and empowering integral part of cultural projects.

Further development is needed of validated and specifically adapted instruments for evaluation during the complete life cycle of cultural projects. Cyclical studies that track how impacts emerging during the creation and end-of-life phases correlate with the ultimate success of the event and its broader cultural resonance would be of great value.

Future Challenges

In order to consolidate the use and application of cultural sustainability assessment methods, it is essential to work on their standardization. The measurement of cultural value has a qualitative and sometimes subjective nature, which can make it difficult to establish universally applicable indicators. In addition, it is noted that cultural impact can manifest itself over long periods of time, which also makes it difficult to collect. These intangible outcomes, such as changes in identity or memory, pose a challenge to standardized and universal quantification using conventional life-cycle assessment tools.

Another challenge is the inclusion of participatory evaluation models. In order to achieve a holistic approach to the whole life cycle, it is essential to incorporate the perspective of new stakeholders beyond the public. Museographers, curators or technical staff, among other stakeholders, can bring a totally innovative perspective to understanding the cultural impact. This requires balancing the expectations of all stakeholders and ensuring methodological coherence to be applied.

Furthermore, integrating cultural impact into existing assessment models based on the three pillars of sustainability (environmental, social and economic) requires methodological innovation. In particular, it is necessary to harmonize units of measurement as well as data collection techniques. On the other hand, advocating the application of comparative frameworks allows for an effective assessment of cultural value, but also represents a major challenge to be developed in the future.

References

Biedermann, A., Muñoz, N., Santolaya, J. L., & Valero, J. I. (2022). Sustainability improvement in complex systems composed of products and services. *International Journal of Life Cycle Assessment, 27*(1), 98–121. https://doi.org/10.1007/s11367-021-02014-9

Biedermann, A. M., López, N. M., Lapesa, I. R., Pérez, F. J. G., & Sáenz, J. L. S. (2023). Sustainable services planning. Methods supporting the design of cultural exhibitions. *Heliyon, 9*(9). https://doi.org/10.1016/j.heliyon.2023.e19866

Biedermann, A. M., Muñoz López, N., Santolaya Sáenz, J. L., Asión-Suñer, L., & Galán Pérez, F. J. (2024). Methodological framework for integrating cultural impact in sustainability assessments of cultural events. *Sustainability, 16*(16), 6893. https://doi.org/10.3390/su16166893

Birtch, J. (2024). *A culture of sustainability: The long journey home*. Eliva Press.

Blumenthal, A. R., & Wittlin, A. S. (1971). Museums: In search of a usable future. *Art Journal, 31*(1), 94. https://doi.org/10.2307/775643

Booth-Kurpnieks, C., Dunford, M., & Knell, J. (2023). Impact & Insight Toolkit: Dimensions—The history and development. *Counting What Counts*. https://impactandinsight.co.uk/app/uploads/2023/11/CWC-Dimensions-History-NOV-2023-Adobe.pdf

British Council. (2020). *The missing pillar*. pp. 379–396.

Brocchi, D. (2008). *The cultural dimension of sustainability cultures of sustainability cultural strategies of sustainability*, (1975), pp. 270–271.

Burns, T. R., Corte, U., & Machado, N. (2015). The sociology of creativity: PART II: Applications: The socio-cultural contexts and conditions of the production of novelty. *Human Systems Management, 34*(4), 263–286. https://doi.org/10.3233/HSM-150850

CDN, Cultural Development Network (2025). https://culturaldevelopment.net.au/takso/

Ellen MacArthur Foundation. (2013). *Towards the Circular Economy: An economic and business rationale for an accelerated transition*.

Fairclough, G., Harrison, R., Jameson, J. H., & Schofield, J. (2008). *The heritage reader*. Routledge.

Falk, J. H., & Dierking, L. D. (2016). *The museum experience revisited*. Routledge.

Falk, J. H., & Lynn, D. (2008). *Learning from museums: Visitors experiences and the making of meaning*. AltaMira Press.

George, R. (2024) Cultural impact of sustainability, Exploiting Indigenous Practices. Bedlam Buster.

Hawkes, J. (2001, May). *The fourth pillar of sustainability: Culture's essential role in public planning* (p. 69). Common Ground Publishing Pty Ltd in Association with the Cultural Development Network (Vic).

Jelinčić, D. A. (2021). Indicators for cultural and creative industries' Impact Assessment on Cultural Heritage and Tourism. *Sustainability, 13*(14). https://doi.org/10.3390/su13147732

Jelinčić, D. A. (2022). *Economic sustainability of culture and cultural tourism*. MDPI Books.

KEA European Affairs. (2009). *The impact of culture on creativity*. Retrieved May 2, 2025, from https://keanet.eu/wp-content/uploads/2019/09/impactculturecreativityfull.pdf

Lowenthal, D. (1985). *The past is a foreign country*. Cambridge University Press.

Muñoz, N., Biedermann, A., Santolaya, J. L., Valero, J. I., & Serrano, A. (2021). Sustainability in the design of an itinerant cultural exhibition. Study of two alternatives. *Applied Sciences (Switzerland), 11*(21). https://doi.org/10.3390/app11219863

O'Brien, D., & Oakley, K. (2015). *Cultural value and inequality: A critical literature review*. Arts and Humanities Research Council.

Pascual, J. (2016). Agenda 21 for culture. Sustaining Cultural Development: Unified Systems and New Governance in Cultural Life, 37–42. https://doi.org/10.4324/9781315611594.

Sagnia, B. K. (2004). Framework for cultural impact assessment. *International network for cultural diversity (INCD)*.

Soini, K., & Birkeland, I. (2014). Exploring the scientific discourse on cultural sustainability. *Geoforum, 51*, 213–223.

Soini, K., & Dessein, J. (2016). Culture-sustainability relation: Towards a conceptual framework. *Sustainability, 8*(2), 167.

Stephenson, J. (2023). *Culture and sustainability: Exploring stability and transformation with the cultures framework*. Springer Nature.

Tchermalykh, N. (2024, June 18). Anthropology of art and performance. Oxford Research Encyclopedia of Anthropology. Retrieved June 6, 2025, from https://oxfordre.com/anthropology/view/10.1093/acrefore/9780190854584.001.0001/acrefore-9780190854584-e-644. https://doi.org/10.1093/acrefore/9780190854584.013.644.

Throsby, D. (2017). Culturally sustainable development: Theoretical concept or practical policy instrument? *International Journal of Cultural Policy, 23*(2), 133–147. https://doi.org/10.1080/10286632.2017.1280788

UNESCO. (2019). Culture 2030 Indicators.

Vervoort, J., Smeenk, T., Zamuruieva, I., Reichelt, L. L., van Veldhoven, M., Rutting, L., et al. (2024). 9 dimensions for evaluating how art and creative practice stimulate societal transformations. *Ecology and Society, 29*(1), 29. https://doi.org/10.5751/ES-14739-290129

Wallace, L. (2024). Demonstrating the impact of arts, culture and events. CultureCounts. Retrieved June 6, 2025, from https://culturecounts.cc/blog/demonstrating-the-impact-of-arts-culture-and-events

Worts, D. (2003). On the Brink of Irrelevance?. In: Xanthoudaki, M., Tickle, L., Sekules, V. (Eds.), *Researching Visual Arts Education in Museums and Galleries. Landscapes: The Arts, Aesthetics, and Education, vol 2*. Springer, Dordrecht. https://doi.org/10.1007/978-94-010-0043-7_15

Worts, D. (2004). Museums in search sustainable of future. *Alberta Museums Review.*, 40–57.

Worts, D. (2011). Culture and museums in the winds of change: The need for cultural indicators. *Culture and Local Governance, 3*(1), 117–132.

Worts, D., & Dal Santo, R. (2022). The inside-outside model animating the muses for cultural transformation amid the climate crisis. In N. Borelli, P. Davis, & P. del Santo (Eds.), *Ecomuseums and climate change*. Ledizione Press.

Yan, W.-J., & Liu, S.-T. (2023). Creative economy and sustainable development: Shaping flexible cultural governance model for creativity. *Sustainability, 15*(5). https://doi.org/10.3390/su15054353

Zhang, T. (2024). The impact of observers on peoples behavior. *Lecture notes in education psychology and public Media, 50*(1), 243–248. https://doi.org/10.54254/2753-7048/50/20240957

CHAPTER 5

Cultural Heritage Preservation: Reflections on Sustainability and Cultural Impact

Estrella Sanz Domínguez

Terminological Introduction to the Different Conservation Measures and Actions

The conservation of cultural heritage entails various measures and actions aimed at safeguarding it, which typically affect issues such as the number of cultural assets involved, the objectives and expected results, or, among others, established criteria for undertaking them based on their heritage value (Spanish Association for Standardization and Certification [AENOR], 2024). At the same time, any of these measures and actions should be carried out taking into consideration the dimensions that may shape the heritage identity of the element or elements being addressed, namely: material and technical properties, aesthetics, historical, documentary, scientific, social value, etc. (García Martínez & Díaz-Sierra, 2023; AENOR, 2024). Based on these premises, it would be advisable to conduct a thoughtful analysis of the sustainability and cultural impact of these

E. Sanz Domínguez (✉)
Faculty of Fine Arts, Complutense University of Madrid, Madrid, Spain
e-mail: essanz@ucm.es

A. M. Biedermann, F. J. Galán Pérez (eds.), *Interconnections between Culture and Sustainability*,
https://doi.org/10.1007/978-3-032-21061-6_5

recognized conservation measures and actions, known as *restoration, curative conservation, and preventive conservation*.

In order to understand what aspects the aforementioned conservation actions comprise, we can start from the definitions that international organizations such as International Council of Museums (ICOM) or regulations such as UNE-EN 15898:2020 *Conservation of Cultural Heritage: Main General Terms and Definitions*, and others related to conservation of cultural heritage, assign to these concepts.

In 2008, during the 15th International Conference of the ICOM Committee for Conservation, held in New Delhi, a document was issued that adopted these terms as part of conservation, defining them to facilitate communication among professionals in the field. Thus, it establishes the concept of *restoration* as interventions carried out directly on a stable asset that has suffered a loss of meaning or functionality, the purpose of which will be to promote its appreciation and use, always paying attention to respect for the dimensions considered above. In the case of *curative conservation*, it determines that its objective will be to halt, through direct actions, active and risky deterioration processes or to structurally reinforce an asset or group of assets. Finally, it associates *preventive conservation* with indirect measures and actions, the objective of which will be to prevent or minimize future damage or loss (International Council of Museums [ICOM-CC], 2008). It also points out the possibility that some of these measures and actions may overlap in purpose.

In the case of the definitions established in the UNE-EN 15898:2020 standard for terms associated with conservation measures and actions, we see that they are basically the same, albeit with some qualifications. Regarding the term “restoration,” it incorporates the idea that the actions applied may also be aimed at highlighting the heritage value of the asset; that is, it identifies this issue as a possible criterion for undertaking the intervention. In the case of *curative conservation*, although in line with ICOM-CC, its definition will be somewhat more concise, not referring, for example, to a condition for action determined by the severity of the asset’s state of conservation. Associated with this term, it also speaks of *corrective conservation*, a term that could be interpreted, without the relevant qualification, as equally possible actions within the framework of restoration or *preventive conservation*. Finally, with regard to preventive conservation, it should be noted that in both cases, measures or actions are being discussed aimed at avoiding or minimizing future damage or deterioration, but, in addition, the aforementioned standard mentions the intention of avoiding future invasive interventions (Spanish Association for Standardization and Certification [AENOR], 2020), in a more sustainable line than heritage conservation entails if we think long term and in terms of the cost of the action.

Sustainability and Cultural Impact in Cultural Heritage Conservation Plans

Once we have reviewed this terminology and its associated objectives, we can begin to reflect on the sustainability and cultural impact that these measures and actions can have within the framework of cultural heritage conservation. That is, on the policies and criteria we must adhere to so that any of the aforementioned conservation actions are sustainable over time and responsible with economic, environmental, and social resources. They also contribute to transforming the environment with a lasting positive impact (Spanish Network for Sustainable Development [SDSN], 2022; Vivancos Ramón, 2023).

However, it is true that the concept of conservation itself implicitly implies the idea of sustainability, or at least part of it, by aiming for the permanence of heritage assets over time to make them accessible to society in the short, medium, and long term (AENOR, 2020; ICOM-CC, 2008) and also recognizing their value not only as a material support for an aesthetic or image, but also as a representation of a social and cultural identity. This is the vision of some authors, such as Valerie Magar, when they speak of two facets of sustainability in conservation: one focused on the conservation actions themselves, and another on what conservation itself as an end can contribute to sustainable development in a society in continuous transformation (Magar Meurs, 2016).

When it comes to decision-making, planning, and implementing the various conservation processes or plans, involving professionals from different disciplines, there must always be a consensus based on knowledge of the asset or assets, their use, and their management; within, of course, the established criteria of "best practices" reflected in regulations, charters, etc., but without forgetting the social impact that these actions may entail. There must also be a flow of knowledge and information in both directions, hence the importance of including this factor of social impact in the equation to establish the sustainability of the conservation action undertaken. On the other hand, although the term conservation leads us to think that it is an ongoing, constant, and sustained process over time, the truth is that some of these actions, such as *restoration* or *curative conservation*, end with the delivery of a project in which, at most, some medium/long-term conservation recommendations will be incorporated (Spanish Association for Standardization and Certification [AENOR], 2018)—although without effective supervision, in many cases, of the monitoring and implementation of these recommendations or, more disturbingly, without the certainty that the custodian institution of that cultural asset can guarantee these preventive conservation measures outlined.

Another interesting aspect of conservation is mentioned in the UNE-EN 16835 standard, on the conservation of cultural heritage, decision-making, planning, and implementation of processes, which is to "manage change in a way that supports and enhances the value of cultural heritage" (AENOR, 2018, p. 6). This statement can be interpreted as meaning that we must be able to maintain cultural assets that can evolve due to endogenous and/or exogenous factors, while also contributing to protecting their value and usefulness. This reflects a clear link with the idea of sustainability and the cultural impact of conservation actions. This management of change or evolution of tangible cultural heritage involves actions to ensure its durability, in line with the heritage value of that asset or those cultural assets in a given social context. This can undoubtedly be a determining factor in the sustainability and cultural impact of conservation actions.

Where, then, can we find aspects associated with sustainability and cultural impact in cultural heritage conservation projects (AENOR, 2018)? For example:

- In project communication, assessing the scope that this should have in relation to the project's cultural impact.
- In defining project objectives, when it is requested to include environmental risk issues and social considerations that may affect, for example, usage patterns.
- In the research and assessment phase, when establishing the significance or value of the heritage to be addressed. We already mentioned at the beginning of the text how this heritage interest can determine a specific type of conservation action. Also, within this assessment phase, the environmental information and assessment collected will be associated with the subsequent environmental and economic impact of our action.

- In assessing and managing project risk throughout its development, as well as its future implications, including environmental, tourism, economic, and social impact. This assessment could determine some of the conservation measures and actions most aimed at stabilization, aesthetic restoration, changes associated with management or use, motivated by accessibility, carbon footprint, or resources involved.
- In quality management, to ensure and control that all established procedures based on sustainability criteria are carried out.
- Finally, as previously mentioned, every conservation project or plan, once completed, must include guidelines for use, maintenance, and management to make it truly sustainable.

Conservations and Associated Polynomials from a Sustainable Perspective

The identified conservation actions—*restoration*, *curative conservation*, and *preventive conservation*—could also be studied as a trinomial composed of a series of constants and variables that can determine their sustainability (Fig. 5.1).

In a first approach, we could establish the predictable constants of these monomials as the cultural asset itself and its material and formal nature, as well as its geographic location, initially ignoring the evolution of its state of conservation and the effects of climate change associated with the geographic/climatic zone. The variables would be its use and management and the evolution of the social context to which it is linked.

Restoration	**Preventive Conservation**
Curative Preservation	**CONSERVATION** **Predictable constants** Material nature of the asset Geographic location **Variables** Use and management Social context

Fig. 5.1 Simplified graphical representation of conservation actions in the form of a trinomial. (Source: Estrella Sanz)

Some of the issues to consider when establishing sustainable conservation strategies, taking into account the variable of use and management, associated with the *preventive conservation* monomial, would begin with the building envelope (Spanish Association for Standardization and Certification [AENOR], 2019), i.e., the building that houses the heritage collections. These are:

- Projected energy use related to air conditioning (determination of tolerance in annual cycles based on the material and technical nature of the collections), lighting, and other administrative and technical energy consumption processes primarily associated with environmental and economic impact, although with a reflection of social responsibility. The objective to be pursued would be the energy efficiency of the processes and the trend toward low- or near-zero--energy consumption envelopes.

  Here, the constants related to the material nature (sensitivity of the collections) and geographic location would be a determining factor in the projected energy use, in the requirements of the needs and design of the envelope, as well as in the potential energy consumption due to travel by potential users (AENOR, 2019).
- Fire safety related, above all, to the choice of extinguishing system, which must be compatible with the safety of the parts, effective, environmentally friendly, and cost-effective.

  Regarding some of these systems, it is important to consider, for example, the economic and environmental impact of energy costs in systems based on oxygen reduction or the environmental impact of systems based on halogenated hydrocarbons, extinguished due to their impact on the ozone layer and replaced, in some cases, by inert gas systems of nitrogen, argon, and nitrogen dioxide, halocarbon agents (National Institute for Safety and Health at Work, 2013), or water mist, whose system reduces water consumption (National Institute for Safety and Health at Work, 2004).
- Regarding administrative management, the economic cost of the resources used to finance the cultural entity's own activities and projects must be considered (Ateca Amestoy, 2022): collecting, research, conservation, exhibition, etc., in addition to those associated with building maintenance to enable the aforementioned activities. Administrative management within a sustainable conservation strategy must also implement a procurement policy that balances

economic, environmental, and social impact, considering the availability of present and future resources and spaces.

Once the policy establishing a framework for action, the purpose, and objectives of the institution responsible for the conservation of cultural assets has been defined, it will be necessary to conduct impact assessments of these activities, assessing their transformative power (Spanish Association for Standardization and Certification [AENOR], 2024). Similarly, and taking into account the "heritage interest" factor in prioritizing potential conservation actions, risk assessment is another point to consider. This assessment will allow for the identification, analysis, and evaluation of potential risks to the collections, assessing the likelihood of occurrence and the impact on all aspects considered in the asset or group of assets. This implies that institutions that are custodians of heritage collections have the duty and responsibility to conduct assessments and reviews of the heritage value of their collections, as indicated in the UNE-EN 17820 (2024) standard, specifications for the management of movable cultural heritage collections, with the aim of achieving optimal risk management and adequate emergency planning.

The aforementioned standard also emphasizes, as a collections management principle, sustainability associated with the environmental impact (energy consumption and carbon footprint) in the development of all activities and circumstances of use of the collections: storage, loans, deposits, exhibition, etc. (AENOR, 2024).

If the conclusion has been reached that a cultural asset requires intervention, this must be the result of an intensive prior study, in which the examinations carried out establish a clear diagnosis that leads to an appropriate and well-founded proposal. This prior diagnosis and analysis will also make it possible to establish subsequent preventive conservation protocols that minimize deterioration, prolonging and even avoiding future interventions (Vendrell Saz et al., 2023). That said, regarding the monomials *curative conservation* and *restoration*, there are relevant aspects associated with sustainability and cultural impact that should be considered, of course always under a deontological commitment linked to professional ethics. First, we could refer to the contracting processes for these services (Spanish Association for Standardization and Certification [AENOR], 2021). Following the usual order of the stages of the contracting procedure, we can extract some of the most relevant issues. Namely:

- Regarding the justification of the need for the service, the heritage value of the intervention should be clearly established to help determine the cultural impact of the intervention.

  Similarly, and with the intention of ensuring the sustainability of the object under contract, it would be necessary to describe the circumstances of use that the object will have once restored, specifying the conditions of the environment in which it will be located. If there is no guarantee of adequate conservation in the short, medium, and long term after the intervention, the project's sustainability model would be compromised.
- The choice of contracting system may be determined, among other factors, by issues associated with economic impact, although other conservation specifications, linked to sustainability during project development, should also be addressed and assessed.
- The drafting phase of the conservation-restoration specifications is a key point in determining the sustainability and cultural impact model of the project to be contracted. Some of the influential aspects to highlight include the precision of the project objectives; a technical report consistent with the principle of sustainability, as well as the details of the work specifications related to conservation and restoration; the incorporation of sustainability into packaging and transport conditions; and the development of citizen participation, communication, and dissemination measures to foster cultural impact and/or tourism promotion (Alonso Campanero et al., 2022).

  Administrative content may include, within the criteria for selecting and evaluating bids, valuable aspects related to sustainability and cultural impact, as well as mandatory compliance with related regulations.
- The evaluation and selection matrix for proposals to carry out the conservation-restoration project may incorporate valuable criteria related to the sustainability of the proposal.

When a conservation-restoration project is proposed, the sustainability model must be assessed from a comprehensive perspective that is not limited exclusively to the use of more ecological materials, the reduction of energy impact, or waste (Vendrell Saz et al., 2023). Thus, the sustainable concept must be incorporated from the contracting process itself, taking into account the economic, social, environmental, and cultural impact, and after the aforementioned prior examination and diagnosis, ignoring the idea of conserving the asset exclusively in its material, formal, or aesthetic essence (An et al., 2022).

For example, in the case of Spain, *Law 9/2017, of November 8, on Public Sector Contracts, which transposes into Spanish law Directives 2014/23/EU and 2014/24/EU of the European Parliament and of the Council of February 26, 2014*, establishes in its article 99.1 the possibility of incorporating, in the object of the contracts, considerations regarding technological, social, or environmental innovation with the aim of improving the efficiency and sustainability of the goods or services to be contracted. At the same time, *Law 7/2021, of May 20, on climate change and energy transition*, in its article 31, and in accordance with the aforementioned *Law 9/2017, of November 8*, points out issues such as the incorporation into the corresponding public procurement specifications of weighted award criteria related to the environment and energy sustainability. This, in turn, implies introducing into the specific technical specifications, without restricting contracting, criteria set forth in *Royal Decree 163/2014, of March 14, which creates the registry of carbon footprint, carbon dioxide offsetting, and carbon dioxide absorption projects.*

Conservation and Sustainable Criteria in Cultural Heritage

The incorporation of sustainable criteria in both curative conservation and restoration, as well as in preventive conservation, must begin, as we have explained, with the exploration of areas of sustainability within these tasks that favor effective actions. As mentioned above, perhaps the simplest trend is to think of sustainable interventions associated, almost exclusively, with aspects linked to the selection of materials and conservation--restoration procedures for both movable and immovable properties, with the particularities of the latter (Vivancos Ramón, 2023). However, if we intend to achieve these sustainability objectives, in addition to the usual items that form the basis of these actions, such as minimal intervention, reversibility, legibility, stability, documentation, etc. (Muñoz Viñas, 2003), we must add others that require prior reflection before the intervention proposal, that is, during the decision-making process, and which have to do with the so-called heritage impact assessment (HIA) (Ashrafi et al., 2022)—aimed, in this case, at their prevention, mitigation, or valorization, through a mechanism that identifies, predicts, and assesses the consequences of an action on cultural heritage (García Martínez & Díaz-Sierra, 2023).

In 2011, the International Council on Monuments and Sites published a guide on heritage impact assessments for World Heritage properties, which incorporates some notable aspects that can be extrapolated to other cultural heritage sites that are susceptible to intervention. Below, we highlight some of the most relevant issues that could be applied to conservation criteria for greater sustainability (International Council on Monuments and Sites [ICOMOS], 2011):

- Regarding the guidelines for impact assessments of cultural heritage subject to any proposal—in our case, for protection, conservation, or restoration—a key initial point is noted: the importance of conducting a preliminary study of cultural and/or natural heritage attributes, in parallel with an environmental impact assessment (EIA) that such action might have. It is true that, from the perspective of this document, when referring to World Heritage, it indicates the need for a broader view regarding the competition between the conservation requirements of that heritage and the uses and developments to which it might be exposed due to its circumstances. Perhaps this would be more applicable and relevant to this study if we were discussing conservation-restoration interventions on immovable heritage assets.
- Another interesting issue introduced in this document, which could be extrapolated to the topic under review, is the approach to a heritage and/or environmental impact study, or even the conservation--restoration project itself, under an iterative process. The advantages of this process lie in that it allows for continuous review of criteria and proposals until a more satisfactory and appropriate solution is reached. Its objective is to reduce risk, which is fundamental in any approach to cultural heritage conservation, and to manage it efficiently, which is essential to ensure sustainability.
- Regarding the contents indicated and that should be reflected in the heritage impact assessment report, a series of points also appear that would undoubtedly contribute to establishing more sustainable and appropriate conservation criteria, taking into account their potential social, economic, and environmental impact (Ashrafi et al., 2022). Thus, it is considered relevant to present in this report aspects related to its value, integrity, authenticity, state of conservation, context in

which it is inserted, and possible interrelationships. And it is this last issue, the possible interrelationships, that we want to highlight—that is, the importance of knowing the real or conventional connection with the user, or the use made of that cultural heritage, which must be conserved, in order to achieve sustainability criteria more aligned with the different dimensions of what we consider sustainable.

Another point mentioned as part of the required content in the heritage impact report, and applicable to sustainable conservation criteria, is that it establishes the heritage benefits likely to arise as a result of one or another conservation proposal.

- Finally, it is important to note the importance of conducting a scoping study of the action to be carried out, in order to tailor this study as closely as possible to its purpose.

Ultimately, the goal will be to find balanced solutions between accepted and standardized conservation criteria and contemporary sustainability criteria (International Council on Monuments and Sites [ICOMOS], 2017). How can this be achieved? For example, by promoting research for the development of new materials or procedures that directly or indirectly affect cultural heritage; assessing the cultural and environmental impact of these interventions; or promoting training programs for technicians involved in heritage conservation, and education, information, and dissemination of sustainable practices for its users, in which all stakeholders responsible for the conservation and protection of cultural heritage must be involved (Alonso Campanero et al., 2022).

References

Alonso Campanero, J. A., Martínez Cuart, I., Núñez García, M., Aguilar Benavides, L. J., & Ulargui Aparicio, C. (2022). *Guía de criterios de sostenibilidad en la conservación y restauración del patrimonio cultural inmueble (Guide to sustainability criteria in the conservation and restoration of immovable cultural heritage)*. Consejería de Cultura, Turismo y Deporte, Dirección General de Patrimonio Cultural, Subdirección General de Patrimonio Histórico.

An, L., Song, C., Zhang, Q., & Bohnett, E. (2022). *Conservation effectiveness and concurrent green initiatives*. Routledge, Taylor & Francis Group.

Ashrafi, B., Neugebauer, C., & Kloos, M. (2022). A conceptual framework for heritage impact assessment: A review and perspective. *Sustainability, 14*, 27. https://doi.org/10.3390/su14010027

Asociación Española de Normalización y Certificación (Spanish Association for Standardization and Certification). (2018). *Conservación del Patrimonio Cultural: Proceso de conservación. Toma de decisiones, planificación e implementación (Conservation of cultural heritage: Conservation process. Decision making, planning and implementation)*. UNE-EN 16853. https://www.une.org/encuentra-tu-norma/busca-tu-norma/norma?c=N0060734

Asociación Española de Normalización y Certificación (Spanish Association for Standardization and Certification). (2019). *Conservación del Patrimonio Cultural: Especificaciones para el emplazamiento, construcción y modificación de edificios o salas destinadas al almacenamiento o utilización de colecciones del patrimonio (Conservation of cultural heritage: Specifications for location, construction and modification of buildings or rooms intended for the storage or use of heritage collections)*. UNE-EN 16893. https://www.une.org/encuentra-tu--norma/busca-tu-norma/norma?c=N0061358

Asociación Española de Normalización y Certificación (Spanish Association for Standardization and Certification). (2020). *Conservación del Patrimonio Cultural: Principales términos generales y definiciones (Conservation of cultural heritage: Main general terms and definitions)*. UNE-EN 15898. https://www.une.org/encuentra-tu-norma/busca-tu-norma/norma?c=N0064615

Asociación Española de Normalización y Certificación (Spanish Association for Standardization and Certification). (2021). *Conservación del Patrimonio Cultural: Contratación de servicios de conservación y obras para el patrimonio cultural (Cultural heritage conservation: Contracting of conservation services and works for cultural heritage)*. UNE-EN 17429. https://www.une.org/encuentra-tu-norma/busca-tu-norma/norma?c=N0065385

Asociación Española de Normalización y Certificación (Spanish Association for Standardization and Certification). (2024). *Conservación del Patrimonio Cultural: Especificaciones para la gestión de las colecciones del patrimonio cultural mueble (Conservation of cultural heritage: Specifications for the management of movable cultural heritage collections)*. UNE-EN 17820. https://www.une.org/encuentra-tu-norma/busca-tu-norma/norma?c=N0072402

Ateca Amestoy, V. (2022). Análisis económico para entender los museos y sus misiones *(Economic analysis to understand museums and their missions). RdM, 84*, 65–74.

García Martínez, E., & Díaz-Sierra, R. (2023). El patrimonio cultural en las evaluaciones de impacto ambiental (Cultural heritage in environmental impact assessments). *Ge-Conservacion, 23*(1), 18–27. https://doi.org/10.37558/gec.v23i1.1148

ICOM-CC. (2008). *Terminología para definir la conservación del patrimonio cultural tangible (Cultural heritage in environmental impact assessments. Terminology to define the conservation of tangible cultural heritage)*. 15ª

Conferencia Trienal, New Delhi, 22–26 de septiembre. https://ge-iic.com/files/Cartasydocumentos/2008_Terminologia_ICOM.pdf

ICOMOS. (2011). *Guidance on heritage impact assessments for cultural world heritage properties*. International Council on Monuments and Sites. Traducción 2016. https://icomos.es/wpcontent/uploads/2021/01/Estudio-de--Impacto-Patrimonial.pdf

ICOMOS. (2017). *Approaches to the conservation of twentieth-century cultural heritage. Madrid – New Delhi Document*. International Scientific Committee on Twentieth Century Heritage. https://openarchive.icomos.org/id/eprint/2695/1/MNDD_SPANISH.pdf

Instituto Nacional de Seguridad e Higiene en el Trabajo. (2004). *NTP 666 Sustitutos y alternativas para los halones en extinción (NTP 666 Substitutes and alternatives for halons in fire extinguishing)*. https://www.insst.es/documents/94886/326775/ntp_666.pdf/595d6f64-580d-43db-9998-b59fedb4ffb7?version=1.2&t=1692772659377

Instituto Nacional de Seguridad e Higiene en el Trabajo. (2013). *NTP 975 Instalaciones de extinción automática con agentes extintores gaseosos (NTP 975 Automatic extinguishing systems with gaseous extinguishing agents)*. https://www.insst.es/documents/94886/326775/ntp-975w.pdf/fe5990cb-704a-4990-ba3d-2159b88e8b63?version=1.1&t=1687296950279

Magar Meurs, V. (2016). Sostenibilidad y conservación del patrimonio cultural. *CR. Conservación Y Restauración, 10*, 8–14. https://revistas.inah.gob.mx/index.php/cr/article/view/11920

Muñoz Viñas, S. (2003). *Teoría contemporánea de la restauración (Contemporary theory of restoration)*. Editorial Síntesis.

Red Española para el Desarrollo Sostenible. (2022). *El impacto de la cultura en los objetivos de desarrollo sostenible* (The impact of culture on the sustainable development goals). https://culturasostenible.org/wp/wp-content/uploads/2022/11/El-impacto-de-la-cultura-en-los-ODS.-Documento-de-conclusiones-2022-.pdf

Vendrell Saz, M., Juanes Barber, D., & Gómez Morón, A. (2023). Sostenibilidad y patrimonio histórico: De la innovación tecnológica a la tradición en conservación y restauración (Sustainability and historical heritage: From technological innovation to tradition in conservation and restoration). *Revista PH, 110*, 34–49. https://doi.org/10.33349/2023.110.5400

Vivancos Ramón, V. (2023). ¿Qué significa la sostenibilidad en conservación-restauración de bienes culturales? *Revista PH, 110*, 294–295. https://doi.org/10.33349/2023.110.5397

CHAPTER 6

Cultural Dimension and Gender Perspective in Sustainability Analysis of Retail Spaces

Aranzazu Fernández-Vázquez and Anna Maria Biedermann

Introduction

Within the built environment of cities, retail spaces are small portions where social interactions are developed to provide products and/or services to the general public, usually in exchange for money. The commercial nature of the activity inside them does not exclude other dynamics occurring simultaneously, even leaving aside the very social and cultural purpose that some of them may have (i.e. libraries, galleries, and music shops).

These interactions and dynamics link commercial spaces directly with the definition of culture by Raymond Williams as a system that maintains,

A. Fernández-Vázquez (✉) • A. M. Biedermann
Department of Design and Manufacturing Engineering, University of Zaragoza, Zaragoza, Spain

OAAEP Research Group, Funded by the Government of Aragón with FEDER Funds, Zaragoza, Spain
e-mail: aranfer@unizar.es; anna@unizar.es

A. M. Biedermann, F. J. Galán Pérez (eds.), *Interconnections between Culture and Sustainability*,
https://doi.org/10.1007/978-3-032-21061-6_6

communicates, and reproduces the characteristics of a society and allows for people to participate in it. Commercial spaces are social and cultural environments where individuals and groups construct and negotiate identities, express values, adopt lifestyles, and engage in consumption rituals that are laden with meaning (Byun & Oh, 2025). These are "public places that offer opportunities for daily engagement," often functioning as vital nodes within the urban and social fabric.

As consumers increasingly seek cultural and entertainment experiences in commercial spaces, the ability of these spaces to remain sustainable will inherently depend on how well they manage, nurture, and foster this cultural dimension. In this new paradigm, ignoring cultural sustainability would mean overlooking a key driver of their own transformation and appeal. To avoid this, cultural sustainability has to be incorporated to the core business strategy as a foundational pillar of future viability, going beyond mere "corporate social responsibility."

Shopping habits, preferences for certain products or brands, and the decisions consumers make are deeply shaped by their cultural contexts, value systems, traditions, and the meanings they attribute to objects and services. This way, the act of consuming becomes an intrinsically cultural practice instead a mere acquisition of goods. As early as 1995, the World Commission on Culture and Development (WCCD) emphasized that "perceived needs are socially and culturally determined" and that sustainable development requires "the promotion of values that encourage consumption standards within the bounds of ecological possibility and to which everyone can reasonably aspire" (Kangas et al., 2017). Therefore, retail spaces understood as cultural places must seek to improve their sustainability performance by considering this dimension in addition to the environmental, economic, and social ones.

The effort to reconceptualise commercial spaces through culture and entertainment is not a superficial trend, but a response to deeper cultural shifts in contemporary society, such as the search for meaningful experiences, the construction of identity through consumption and leisure, on top of a response to consumers' growing appreciation for values such as authenticity and connection. Therefore, cultural sustainability is no longer optional or a cosmetic add-on for the retail sector: it's a critical factor for its relevance, competitiveness, and legitimacy.

Culture and Commerce Coexistence

Commercial spaces such as markets have always been meeting places in which people connect, take part in activities, interact, get information, and communicate with each other. Historically, this has been especially true for women at a time when there were few opportunities to leave the closed environment of their home. Thus, by going to the market they had the chance to be part of the community and to socialize while attending to their duties. In dispersed cities, such as the typical US suburbia, malls and shopping centres have also been for many citizens the real public space to meet, as it has been shown on many occasions in films and literature.

The coexistence of culture and commerce can be observed in public spaces such as Jemaa el-Fnaa Square in Marrakech. This emblematic site showcases a natural symbiosis between economic and cultural functions, so harmoniously intertwined that UNESCO inscribed it in 2008 on the Representative List of the Intangible Cultural Heritage of Humanity (UNESCO, 2008). The balance between commercial and cultural offer is outstanding, both expressed through a remarkable diversity that give the square its vibrant, dynamic, and enriching character. The deep-rooted tradition, in organic coexistence with the artisanal nature of most products, makes visiting the square a truly unique experience. Attempts to replicate this coexistence in privately owned shopping centres often fail, as major brands tend to instrumentalize culture as a mere marketing tool to increase sales. These efforts frequently fall short in generating real cultural value or in creating meaningful experiences for communities (Zukin, 2010).

A crucial factor is the presence of *authenticity*, both in terms of the public nature of the space and in terms of the freedom of cultural expression. In many commercial settings, this authenticity is only superficial: the space remains under private control, limiting spontaneous or non-commercial cultural expressions, which are sometimes even censored (Harvey, 2012).

Nevertheless, there are positive examples. In Poznań, Poland, a shopping centre known as Stary Browar, located in a revitalized former brewery in the city centre, not only managed to preserve part of the industrial heritage of the place but also dedicates substantial indoor and outdoor areas to contemporary art galleries and public exhibitions. These spaces are well-funded and offer an ambitious cultural programme that includes free outdoor cinema during the summer, multiple festivals focused on art and design, and other cultural events (Szulce & Januszewski, 2015). However, such examples remain the exception rather than the rule.

In the Mediterranean basin, retail environments, particularly those located at street level or in basement spaces, often incorporate archaeological remains directly into their architectural design. These vestiges are sometimes presented with full documentation, thereby narrating the site's historical evolution. In this way, the store itself becomes not only a showcase for its merchandise but also an embodiment of local identity (Ashworth & Tunbridge, 2000). The design of the physical space, both exterior and interior can engage in a dialogue with history. It can also be achieved in new construction if it is sensitive to the context, materials, and aesthetics of the local environment.

Moreover, through intentional spatial design, a retail environment can transcend its purely transactional role to become a true "third place": a neutral, welcoming setting where community members gather, converse, and build social ties (Oldenburg, 1999). To achieve this, the layout must include clearly defined common areas such as reading nooks, flexible seating clusters, or mini plazas, openly accessible to all visitors. Equally important is an ongoing programme of community-centred activities—workshops, panel discussions, pop-up performances, and maker events—that animate these shared zones and encourage meaningful interaction. By integrating both thoughtful zoning and curated programming, a retail space can strengthen local social capital and foster a deeper sense of cohesion and belonging among its patrons (Putnam, 2000). Examples of those spaces can be found in many cities, such as the Social Centre and Library La pantera Rossa in Zaragoza (Spain) or Cafe con Libros in Brooklyn, New York (USA).

Furthermore, the preservation of both tangible and intangible heritage can also be fostered through the support and promotion of locally produced goods crafted using traditional artisanal techniques. By collaborating with community artisans and highlighting their craftsmanship within the retail space, stores not only sustain age-old production methods but also transmit cultural knowledge and skills to new generations (Smith & Akagawa, 2009).

A final question that arises and deserves further exploration is how digitalization and the rise of e-commerce are reshaping this cultural dimension of physical commercial spaces. Does the need for physical spaces to offer unique and irreplaceable cultural experiences intensify as a counterpoint to online commerce, or are we witnessing a migration and reinvention of these cultural dynamics within digital and omnichannel environments (Karri et al., 2025)?

Definition of Cultural Sustainability in the Context of Retail Spaces

As mentioned, commercial spaces have been integrated into the landscape of cities as part of their built environment, reflecting values related to their history, traditions, and economy, thus forming part of a city's culture. This definition of culture could not, however, be applied to other commercial spaces, such as malls or shopping centres, which do not reflect these values in the same way. Newer and wider conceptions of what is culture, adjusted to the emerging trends in literature and policy, have to be taken into account to value the cultural component of these spaces.

There has been an evolution of the definition of cultural sustainability, parallel to that of social sustainability, from a more traditional view focused on tangible cultural heritage and basic social needs to a view that also includes immaterial aspects, encompassing both tangible and intangible values (Axelsson et al., 2013). Cultural sustainability according to this new conception shall refer to the preservation of the system of values, practices, and symbols of identity, considering their evolution and permanent updating, as well as diversity, creativity, and innovation (Cantar et al., 2024).

According to this wider definition of culture, commercial spaces must form part of the set of players in the search for cultural sustainability in cities, as they can promote actions and reflect values to help maintain and adapt their cultural heritage to people's needs, or host events, shows, and exhibitions that bring community members together to share experiences.

It is essential to clearly define what is meant by "*cultural sustainability*" in this specific context to really integrate the cultural dimension into the sustainability analysis of commercial spaces. Although several conceptualizations can be found in academic literature, they can be grouped into two interconnected but distinct approaches (Järvelä, 2023):

The Sustainability *of* Culture: This approach focuses on the preservation and continuity of cultural elements themselves, safeguarding tangible cultural heritage (such as historic buildings, artefacts, and cultural landscapes) and intangible heritage (including oral traditions, social practices, rituals, and artisanal knowledge); promoting the diversity of cultural expressions; supporting the vitality of contemporary artistic and creative practices; and protecting and fostering the cultural identities of different communities. This perspective aims to ensure that cultural richness and diversity can endure and be passed on to future generations. In this case, what can be evaluated is whether retail spaces and the activities they organize, or the experiences they offer to customers, serve to preserve and disseminate heritage, whether tangible or intangible, and how they can generate cultural impact, either at the individual level or for communities.

Sustainability *through* Culture: This second approach regards culture as a fundamental resource and active agent in achieving the broader goals of sustainable development. From this perspective, cultural values, worldviews, traditional knowledge, narratives, arts, and creative practices can inform, inspire, and facilitate the transition towards more sustainable societies in ecological, social, and economic dimensions. Here, culture is understood as "*a core set of practices and beliefs that can support or inspire the necessary social transition towards more sustainable living*." This means that culture is understood as a vehicle for content and values related to sustainability (Bălan, 2020).

Applying these definitions to the retail sector, a culturally sustainable commercial space should then function on two complementary levels. On the one hand, it should respect, value, and actively contribute to the cultural vitality of its immediate environment. This entails recognizing and celebrating local identity, promoting cultural diversity in its offerings and representations, and being sensitive to the heritage of the place in which it is located. On the other hand, it should leverage its considerable influence and multiple platforms (both physical and digital) to educate, inspire, and enable consumers to adopt consumption practices that are not only environmentally and socially responsible, but also culturally enriching and meaningful. As has been emphasized, "*culture is fundamental to the possibility of a sustainable society*," and commercial spaces, as central nodes of social and economic life, have an unavoidable role in this equation.

This dual function implies a constant dialectical role for the commercial space: it must act simultaneously as a "*good host*" of local culture and as an "*active agent*" in promoting a culture of sustainability. Being a good host means being context-sensitive, avoiding the imposition of foreign cultural models, resisting homogenisation, and steering clear of the superficial exploitation of local cultural elements. Being an active agent, in turn, involves taking responsibility and recognizing the potential to positively influence consumer values and behaviours, encouraging a more reflective, critical, and sustainability-aligned culture of consumption.

This dual nature is inherently dynamic and can give rise to tensions that must be managed creatively and ethically. A commercial space's ability to navigate this duality is closely tied to its business model and to how it defines the "value" it seeks to generate, transcending short-term economic profit in favour of a long-term vision of shared value creation.

Retail Spaces as Influencers on Customer Behaviour

The act of shopping is performed within a public, socially constructed environment, in which individuals tend to adapt their behaviour in response to perceived social norms and expectations. Psychological and sociological studies have described this phenomenon, often referred to as the Hawthorne Effect (Franke & Kaul, 1978). This behavioural adaptation aligns closely with Erving Goffman's dramaturgical theory of social interaction, which conceptualizes everyday life as a form of theatre in which individuals perform roles depending on the setting, audience, and context (Goffman, 1959).

Retail spaces function in this context not only as transactional venues but as carefully staged environments that influence customer behaviour and decision-making. Their design elements, such as lighting, layout, materials, spatial flow, and ambient cues, serve as subtle but powerful influences that shape the consumer's actions, perceptions, and emotional responses. Moreover, these spatial narratives often convey implicit social expectations, guiding customers towards specific patterns of movement, interaction, and consumption.

Consumer's behaviour and the criteria consumers use to evaluate products are therefore not built up in isolation but deeply conditioned by the aesthetic and symbolic dimensions of the retail environment. Retail design becomes, in this sense, a tool of behavioural influence that not only affects individual preferences but also reinforces broader cultural and social norms related to consumption.

Furthermore, while many researchers have attempted to neutralize social-desirability biases through rigorous double-blind study designs (Hróbjartsson et al., 2012), deliberately harnessing the Hawthorne Effect in retail settings can actually serve as a powerful mechanism to encourage more sustainable and socially responsible consumption (Jackson, 2005). However, the primary challenge would be to reconcile these sustainability objectives with the traditional economic imperatives of retail businesses: retailers must find ways to align environmental goals with profitability metrics, ensuring that green initiatives bolster, rather than undermine, the bottom line.

The path to more sustainable consumption lies on both individual and collective behaviour changes, with retailers uniquely positioned to drive this shift. By moving beyond simply stocking eco-friendly products to embedding sustainability into every aspect of the shopping experience, brands can connect with increasingly conscious consumers while gaining a strategic advantage. Curating offerings, whether fair-trade, locally made, durable, or designed for repair, and highlighting their cultural worth (for example, celebrating traditional crafts or regional seasonal foods) not only supports communities and preserves heritage but also reduces environmental impact.

In both physical and digital spaces, retailers can guide shoppers towards greener choices through dedicated "green trails," storytelling merchandising that reveals each product's origins and positive effects, and well-informed staff as sustainability ambassadors. Choice editing, proactively phasing out high-impact items, or partnering with suppliers to cut unnecessary packaging simplifies decision-making and reinforces commitment when communicated transparently. Finally, by framing sustainable consumption as aspirational and culturally resonant, associating it with values like foresight, authenticity, and community care, while providing clear, relevant information through labels, narratives, workshops, and demonstrations, retailers can inspire meaningful action across diverse audiences.

The impact of these strategies is exponentially greater when they're woven seamlessly into every customer touchpoint and genuinely reflect both the brand's cultural DNA and its audience's aspirations. It's not about isolated initiatives, but about embedding sustainability into the very fabric of the retail experience. The most powerful approaches enable consumers to forge or reinforce a positive cultural identity through their purchases, so turning responsible shopping from an obligation into a meaningful way to express who they are, which value-driven community they belong to, and how they're helping shape a better future.

Cultural elements and products are part of commercial spaces in many ways. Music is ever-present both in individual shops and in common areas of shopping centres, with different musical genres perfectly adapted to the type of product or service provided. Big screens broadcast images and videos of events related to products or product-related characters in sports or music shops or show messages focused on provocation and controversy, using powerful images that address social and political issues.

Retail brands can leverage visual communication to foster intercultural dialogue and reinforce cultural sustainability, but they must do so with genuine community engagement rather than mere "shockvertising." Benetton's campaigns of the 1990s and 2000s, which vividly showcased issues of race, gender, and global conflict, succeeded in broadening public discourse on diversity and tolerance (Incze et al., 2018). However, critics warn that decontextualizing social crises for commercial impact can commodify cultural values and obscure power imbalances, undermining authentic heritage preservation (Tinic, 1997). For cultural sustainability in retail, brands must therefore move beyond provocative imagery to establish transparent partnerships with heritage stakeholders, co-develop educational programmes, and ensure that the stories they tell in their stores genuinely benefit—and are controlled by—the communities whose identities they represent.

Graphic design is perhaps the most evident cultural expression present in commercial spaces through signage, banners, posters, and paramount part of the identity of many brands such as Camper, also serving as a communication tool to showcase the ideology, commitment, and the improvement measures implemented by brands, as it is used to show the environmental commitment of the English brand LUSH. This brand is renowned for using its stores as platforms for activism across a wide range of social and environmental causes: from human rights and animal protection to climate justice and gender equity. Its window displays and interior spaces are often transformed into sites of cultural commentary and mobilization, featuring visually striking campaigns and offering specially designated products whose proceeds support partner organizations.

In the role of commercial spaces as host of cultural products, street artists have executed hand-painted murals in stores, such as Footlocker flagships in Champs Elysées or Barcelona. Luxury brands include works of art in their stores to transmit cultural values and ideas, like the Zimmermann shop in Madrid, which displays works of art by Australian artists as a way of honouring their country of origin, or features pieces created especially for these spaces like Loewe P° de Gracia (Barcelona) store, located in a historical modernist building and presenting works of art such as a bamboo sculpture by the Japanese artist Tanabe Chikuunsai IV, a macramé structure of the Spanish artist Aurelia Muñoz, or the installations of Haegue Yang and Richard Wright.

This way, commercial spaces become part of the cultural environment acting on the one hand as hosts of cultural products and on the other hand as transmitters of values and ideas that reinforce their message by conveying their ideas and objectives.

The Gender Perspective: A Backbone of Cultural Impact in Commercial Spaces

The integration of the cultural dimension as a key factor for the sustainability of commercial spaces would be incomplete and, ultimately, ineffective if the gender perspective is not addressed in a transversal and critical manner. Commercial spaces are not neutral, but arenas where gender norms and roles are constructed, reinforced, and potentially challenged, with profound implications for cultural identity, consumer practices, and social equity.

Commercial spaces have historically played a significant role in gender socialization in many ways. Through their spatial design, the advertising they content, the way they categorize and present products (e.g. the ubiquitous "men's clothing" and "women's clothing" sections, or toy aisles segregated by gender), and the types of services they offer, they have contributed to the naturalization and perpetuation of gender stereotypes. The "gendering" of commercial space influences how individuals learn what is considered "appropriate" or "expected" for their gender, affecting everything from clothing choices to professional aspirations and family roles. A clear example, pointed out by academic literature, is how commercial spaces have often taken for granted that women are the primary shoppers for the household and family, so reinforcing this role through their design and marketing.

"The impact of these socially constructed gender roles extends to the shaping of both individual and collective cultural identity. They can limit options for self-expression, channel desires and aspirations along predefined paths, and generate pressure to conform to often restrictive ideals. Some studies indicate that certain groups, such as young women, may perceive commercial spaces as 'zones of judgment,' where they feel observed and evaluated based on their conformity to gender norms regarding appearance and behaviour" (King & Theocharides-Feldman, 2022).

It is therefore crucial that the analysis of cultural sustainability in retail transcends traditional binary conceptions of gender (male/female). Contemporary society recognizes a much broader diversity of gender identities, including non-binary, genderqueer, gender-fluid, and others across the spectrum (Githapradana et al., 2024). Commercial spaces can be deeply exclusionary, uncomfortable, or even unsafe for those who do not identify within binary categories. Conscious design and management can transform these spaces into affirming and validating environments for all gender identities.

The "gendering" of commercial spaces is not an accidental phenomenon or merely a reflection of "natural" preferences: it is a tangible manifestation of cultural and economic power dynamics that have historically privileged certain identities and behaviours, predominantly those of males and heterosexuals, over others. The persistence of designs and practices that reinforce gender stereotypes, such as the rigid division of product sections, or the lack of offerings that accommodate a diversity of bodies and gender expressions, follows market logics that have been built upon these norms while also contributing to their perpetuation. Moreover, the way commercial spaces address gender directly influences consumer habits.

Questioning and deconstructing these logics (Gqola et al., 2024) is therefore not simply a matter of abstract "inclusion," but an act with profound political and cultural implications. Cultural sustainability, if it is to be genuine, must be inherently critical of these power dynamics. More inclusive spaces that offer a broader diversity of representations and choices, and that do not penalize deviation from stereotypes, can foster more reflective consumption patterns, less driven by the need to conform to restrictive gender norms and more oriented towards genuine self-expression and individual well-being.

A cultural sustainability that aims to be genuine and comprehensive is inseparable from gender equity. Research indicates that women and individuals with diverse gender identities often have different expectations, concerns, and behaviours related to sustainability and consumption. Some studies suggest that women, on average, tend to exhibit higher levels of environmental concern and hold higher expectations regarding companies' sustainability practices (Grunwald et al., 2026). Aspects such as perceived safety, accessibility, comfort, and the sense of belonging in public and commercial spaces are experienced differently depending on gender (Koskela, 1997; Madanipour, 2003). Women and gender minorities may also face greater risks of harassment, or feel more vulnerable in certain environments, and this affects their freedom to move, participate, and enjoy these spaces (Valentine, 1989; Kern, 2020). Ignoring these differences can lead to an incomplete or biased understanding of the cultural impact of a commercial space and risk the effectiveness of its sustainability strategies.

At this point, the concept of "intersectionality" becomes paramount, as gender does not operate in isolation but intersects and interacts with other identity markers such as race, ethnicity, social class, physical or cognitive ability, sexual orientation, and age (Crenshaw, 1991; Hancock, 2007). These intersections shape unique experiences of inclusion, exclusion, privilege, or marginalization. Therefore, an effective gender perspective in the analysis of cultural sustainability must be inherently intersectional, recognizing the complexity and diversity of lived experiences (Kaijser & Kronsell, 2014).

Gender blindness in the design, management, and evaluation of commercial spaces is not only an ethical omission or a social injustice; it also represents a missed opportunity for business and innovation (Greed, 2005; Adams & Tiesdell, 2013). By failing to adequately consider the needs, desires, and perspectives of more than half of the population, and the vast diversity within that group, retailers may be overlooking market niches, missing ideas for new products or services, and failing to build loyalty with significant consumer segments. If, as some studies suggest, women tend to have higher expectations for sustainability and are, in many cases, the primary purchasing decision-makers in households (Dobscha, 1993), then retailers who ignore their perspectives on what constitutes a "culturally sustainable" space or product are missing a strategic opportunity to connect with a key segment.

Similarly, designing for gender diversity, including the specific needs of trans and non-binary individuals, can foster customer loyalty and attract communities that have often been underserved or rendered invisible by the market (Vincent, 2020).

Conclusions and Future Research

This chapter argues for the urgent need to move beyond traditional models of sustainability in the retail sector, which have often neglected the cultural dimension. Culture can not be seen as a decorative element or a niche concern, but rather as a foundational pillar and transformative force that must be systematically integrated into the conception, design, management, and evaluation of commercial spaces. The conceptual evolution towards a "fourth cultural pillar" highlights cultural sustainability as a strategic imperative.

A key guiding thread throughout this proposal is the gender perspective, understood as a critical lens that should permeate all dimensions of cultural sustainability instead of an isolated topic. Since commercial spaces have historically played a role in shaping gender roles, genuine sustainability must aims to deliberately deconstruct them, promoting equity and creating truly inclusive environments for all gender identities.

The transformative practices analysed, ranging from independent stores functioning as vibrant community hubs to global brands attempting to redefine their relationship with consumption, demonstrate that change is possible. The success of these initiatives often lies in their ability to resonate with the deep cultural values of consumers and to offer experiences that go beyond the mere act of transaction.

References

Adams, D., & Tiesdell, S. (2013). *Shaping places: Urban planning, design and development.* Routledge.

Ashworth, G. J., & Tunbridge, J. E. (2000). *The tourist-historic city: Retrospect and prospect of managing the heritage city.* Pergamon.

Axelsson, R., Angelstam, P., Degerman, E., Teitelbaum, S., Andersson, K., Elbakidze, M., & Drotz, M. K. (2013). Social and cultural sustainability: Criteria, indicators, verifier variables for measurement and maps for visualization to support planning. *Ambio, 42*(2), 215–228. https://doi.org/10.1007/s13280-012-0376-0

Bălan, C. (2020). How does retail engage consumers in sustainable consumption? A systematic literature review. *Sustainability, 13*(1), 96.

Byun, H., & Oh, J. (2025). Effect of the socio-environmental characteristics and facilities on the local retail clusters in Seoul, South Korea. *Journal of Asian Architecture and Building Engineering*, 1–17.

Cantar, N. M., Endere, M. L., Mikkelsen, C. A., & Zulaica, M. L. (2024). Indicadores de sustentabilidad sociocultural para el patrimonio urbano. Metodología para una ciudad intermedia Argentina. *Empiria. Revista de Metodología de Ciencias Sociales, 62*, 73–109. https://doi.org/10.5944/empiria.62.2024.42011

Crenshaw, K. (1991). Mapping the margins: Intersectionality, identity politics, and violence against women of color. *Stanford Law Review, 43*(6), 1241–1299. https://doi.org/10.2307/1229039

Dobscha, S. (1993). Women and the environment: Applying ecofeminism to environmentally-related consumption. *Advances in Consumer Research, 20*(1), 36–40.

Franke, R. H., & Kaul, J. D. (1978). The Hawthorne experiments: First statistical interpretation. *American Sociological Review, 43*(5), 623–643.

Githapradana, W., Raharja, G., Ratna, T., & Pebryani, N. (2024). Gender and culture inclusivity in the story of hope fashion collection. *ARCHive-SR, 8*(1), 43–51. https://doi.org/10.21625/archive.v8i1.1046

Goffman, E. (1959). *The presentation of self in everyday life*. Doubleday.

Gqola, P. D., Perera, I., Phadke, S., Shahrokni, N., Zaragocin, S., Satija, S., & Ghosh, A. (2024). Gender and public space. *Gender & Development, 32*(1–2), 1–25. https://doi.org/10.1080/13552074.2024.2376976

Greed, C. (2005). Overcoming the factors inhibiting the mainstreaming of gender into spatial planning policy in the United Kingdom. *Urban Studies, 42*(4), 719–749. http://www.jstor.org/stable/43197283

Grunwald, G., Kara, A., & Spillan, J. (2026). Gender influence on business students' expectations for sustainability transformations in higher education: A cross-national study. *International Journal of Sustainability in Higher Education, 27*(2), 497–519. https://doi.org/10.1108/IJSHE-03-2024-0189

Hancock, A.-M. (2007). When multiplication doesn't equal quick addition: Examining intersectionality as a research paradigm. *Perspectives on Politics, 5*(1), 63–79. https://doi.org/10.1017/S1537592707070065

Harvey, D. (2012). *Rebel cities: From the right to the city to the urban revolution*. Verso Books.

Hróbjartsson, A., Thomsen, A. S., Emanuelsson, F., Tendal, B., Hilden, J., Boutron, I., Ravaud, P., & Brorson, S. (2012). Observer bias in randomised clinical trials with binary outcomes: Systematic review of trials with both blinded and non-blinded outcome assessors. *BMJ, 344*, e1119. https://doi.org/10.1136/bmj.e1119

Incze, C. B., Pocovnicu, A., & Vasilache, S. (2018). Marketing communication analysis of Benetton PR campaigns. *Proceedings of the International Conference on Business Excellence, 12*(1), 457–465. https://doi.org/10.2478/picbe-2018-0041

Jackson, T. (2005). *Motivating sustainable consumption: A review of evidence on consumer behaviour and behavioural change*. Centre for Environmental Strategy, University of Surrey.

Järvelä, M. (2023). Dimensions of cultural sustainability—Local adaptation, adaptive capacity and social resilience. *Frontiers in Political Science, 5*, 1285602.

Kaijser, A., & Kronsell, A. (2014). Climate change through the lens of intersectionality. *Environmental Politics, 23*(3), 417–433. https://doi.org/10.1080/09644016.2013.835203

Kangas, A., Duxbury, N., & De Beukelaer, C. (2017). Introduction: Cultural policies for sustainable development. *International Journal of Cultural Policy, 23*(2), 129–132.

Karri, V., Edubilli, S., Sanapala, G. K., Kumar, S. S., & Mathew, D. (2025). Unveiling the moderating effect of cultural context on sustainability practices in retail: A mediation model of customer experience. *International Journal of Innovative Research and Scientific Studies, 8*(1), 2109–2120.

Kern, L. (2020). *Feminist city: Claiming space in a man-made world*. Verso Books.

King, J., & Theocharides-Feldman, O. (2022). Are girls being designed out of public spaces? https://www.lse.ac.uk/research/research-for-the-world/society/are-girls-being-designed-out-of-public-spaces

Koskela, H. (1997). "Bold Walk and Breakings": Women's spatial confidence versus fear of violence. *Gender, Place & Culture, 4*(3), 301–320. https://doi.org/10.1080/09663699725369

Madanipour, A. (2003). *Public and private spaces of the city*. Routledge.

Oldenburg, R. (1999). *The great good place: Cafes, coffee shops, bookstores, bars, hair salons, and other hangouts at the heart of a community*. Da Capo Press.

Putnam, R. D. (2000). *Bowling alone: The collapse and revival of American community*. Simon & Schuster.

Smith, L., & Akagawa, N. (Eds.). (2009). *Intangible heritage*. Routledge.

Szulce, H., & Januszewski, F. (2015). The importance of prestigious projects to the image and status of a city based on example of Shopping, Arts and Business Center Stary Browar. *Marketing i Rynek, 22*(10).

Tinic, S. A. (1997). United colors and untied meanings: Benetton and the commodification of social issues. *Journal of Communication, 47*(3), 3–25.

UNESCO. (2008). *The Jemaa el-Fna Square*. Intangible Cultural Heritage. https://ich.unesco.org/en/RL/space-of-jemaa-el-fna-00014

Valentine, G. (1989). The geography of women's fear. *Area, 21*(4), 385–390. http://www.jstor.org/stable/20000063

Vincent, B. (2020). *Non-binary genders: Navigating communities, identities, and healthcare*. Policy Press.

Zukin, S. L. (2010). *Naked city: The death and life of authentic urban places*. Oxford University Press.

CHAPTER 7

Technological Ecosystems as Cultural Systems: Designing for Human Meaning and Cultural Sustainability

Teresa Blanco and Eva Tausiet

The Product as a Node: Conceptual Evolution and New Value Criteria

Digitalization has profoundly transformed the technological, economic, and cultural landscape. Driven by advances such as artificial intelligence, the Internet of Things (IoT), and blockchain, this revolution has reshaped both the notion of innovation and the very meaning of "product." In the Knowledge Society, innovation models are no longer based solely on technical progress but also on the cultural resignification of objects and the emergence of new patterns of social behavior (Blanco, 2016; Norman & Verganti, 2014; Santhi & Muthuswamy, 2023).

T. Blanco (✉) • E. Tausiet
Department of Design and Manufacturing Engineering, University of Zaragoza, Zaragoza, Spain

Aragon Institute for Engineering Research (I3A), University of Zaragoza, Zaragoza, Spain
e-mail: tblanco@unizar.es; etausiet@unizar.es

A. M. Biedermann, F. J. Galán Pérez (eds.), *Interconnections between Culture and Sustainability*,
https://doi.org/10.1007/978-3-032-21061-6_7

In the context of IoT, a multitude of devices connect to the cloud and are integrated into new services, particularly within domains such as smart cities—environments whose projected social and economic relevance (Perera et al., 2014; Streitz, 2015) has become a reality. This technological consolidation has given rise to a hybrid or "phygital" reality, in which the physical and digital coexist and constantly interact. Technologies such as wearables, virtual assistants, and smart home systems act as active mediators of everyday experience, shaping how we perceive the body, inhabit spaces, and engage socially (Blanco-Novoa et al., 2020; Omol, 2024). This environment not only transforms technological infrastructure and usage dynamics, but also reshapes the meaning and value we assign to everyday objects.

In parallel, the contemporary product is no longer conceived as a stable, autonomous technical object, and design is no longer limited to solving technical or aesthetic functions (Jordan, 2000; Winby & Mohrman, 2018). Just as a scientific object can be analyzed from multiple disciplinary perspectives (Klein, 1994), today's product requires a multifaceted understanding: as a mechanical, electronic, immaterial, economic, emotional, social, and cultural artifact, embedded in sequences of experience that include brand, communication, interaction, and content (Blanco, 2016; Spooner et al., 2011).

This, in turn, has transformed the criteria of quality and value. Products must be innovative, user-centered, technologically enabled, and socially and environmentally sustainable, while also being aware of their material and energy impact (Blanco, 2016; Vroom & Horvath, 2014). Functionality and usability are no longer differentiators but minimum requirements and added value shifts toward the symbolic dimension: the product's ability to connect with individual aspirations, generate meaning, and respond to shared cultural codes (Spinelli et al., 2020).

In this new framework, design moves away from the isolated object and focuses instead on the connected experience. More than creating products, the goal is to build systems that articulate meaningful relationships between people, technologies, and contexts. Products are no longer autonomous entities but nodes within complex networks that shape how we inhabit, interact, and assign value.

From Individual User to Interconnected Subjects

This shift toward relationality redefines not only the product but also the subject who interacts with it. In this new scenario, the user is no longer an individual figure engaging with an autonomous object, but part of a dynamic network of relationships, where multiple users, devices, and services coexist and interact (Blanco et al., 2026a; Vargo et al., 2024).

This transition has shifted the focus of design, which is no longer directed exclusively at isolated individuals but at collectives and communities that interact simultaneously and in coordinated ways with multiple devices. It opens the door to increasingly collective and distributed forms of use. Relationships between people and products have evolved from one-to-one (1–1) models to many-to-one (n–1) and many-to-many (n–n) configurations, generating shared uses (Blanco, Marco et al., 2026a) and emerging forms of technological appropriation on a global scale.

In this interconnected environment, agency is also redistributed. Decision-making no longer rests exclusively on the human subject but is shared among people, algorithms, sensors, databases, and smart devices. Sociotechnical networks emerge in which objects actively participate in decision-making, meaning production, and the shaping of use (Carvalho & Goodyear, 2018; Cash et al., 2017).

Collectively, these changes redefine both the notion of product and that of the user, shifting the focus from isolated relationships to complex, dynamic, and co-evolving relational systems. The product is understood as an active part of a technological ecosystem whose logic transcends the boundaries of the object and is embedded in interconnected social, symbolic, and technical networks (Blanco, 2016; Bresciani et al., 2021; Ulaga et al., 2013; Vargo et al., 2024).

Technological Ecosystems: Definition and Dynamics

Building on this relational framework, the concept of the technological ecosystem offers an integrative perspective for understanding how these components are organized and connected within dynamic systems. The concept of a technological ecosystem describes contemporary configurations of interaction among technologies, users, infrastructures, services, and social contexts (Anwar & Gill, 2019; Briscoe, 2009). From a technical-functional standpoint, technological ecosystems integrate various interrelated components such as physical devices (sensors, wearables, smartphones), technological infrastructures (IoT networks, cloud platforms, data analytics systems), and users or communities with diverse practices, values, and knowledge (Anwar & Gill, 2019). The integration of these elements enables continuous communication among components, allowing for data synchronization, interoperability, and functional adaptation. This connectivity renders digital ecosystems highly configurable, flexible, and modular environments, capable of reorganizing themselves according to the relationships established among their elements (Petrova & Marinova, 2021).

The success of a technological ecosystem, therefore, depends not only on its technical efficiency but also on its capacity to adapt to changing demands, maintain internal coherence, and sustain value-driven relationships among its elements (García-Holgado & García-Peñalvo, 2018; Manikas & Hansen, 2013; Tsujimoto et al., 2018). In this regard, intelligent infrastructures must incorporate qualities such as modularity, scalability, resilience, and self-organization in their design, to ensure continuous evolution and adaptability (Blanco, 2016), thus, sustainability.

These systems do not operate according to linear or hierarchical logics but through distributed configurations that shift depending on usage, context, and the relationships established by the actors themselves. Accordingly, the focus shifts from isolated products to functional networks, where each node—device, service, or user—acts as an active part of the system.

This approach has been applied across multiple domains: in education, through personalized environments that integrate human and digital agents; in healthcare, via platforms that connect patients, professionals, and technological services; in e-commerce, by linking sellers, consumers, and logistics and payment systems; in smart cities, through the coordination of intelligent urban infrastructures; and in digital cultural ecosystems, where heritage institutions, developers, and users collaborate on the preservation and reinterpretation of cultural heritage (Avcu & Koçoğlu, 2025; Sussan & Acs, 2017). Nevertheless, the deployment of such ecosystems across sectors has often been guided by predominantly technical approaches, limiting the recognition of other types of implications.

Technological Ecosystems as Cultural Environments

The study of digital ecosystems has developed in an accelerated yet fragmented manner. According to Dulsrud and Bygstad (2022), this fragmentation stems from three main factors: their multidisciplinary nature, the diversity of contexts in which they are conceptualized and applied, and their transversal deployment across multiple sectors and social domains.

Traditionally—and understandably—these ecosystems have been defined from a technical standpoint, aligned with the technological developments that made them possible. Technologies have often been introduced vertically, prioritizing functional efficiency over gradual and contextual integration processes (Shin & Park, 2017), which has hindered the incorporation of relevant social and cultural dimensions.

Understanding their full scope requires going beyond technical considerations. The notion of technoculture already pointed to the interdependence between technology and culture in shaping social, affective, and cognitive experiences (Verbeek, 2005). From this perspective, technological ecosystems are not merely infrastructures or services; they also produce meaning, construct norms, and project worldviews (Alsaleh, 2024). Far from being neutral, technologies embody implicit visions of the present and future and actively participate in the production of subjectivities and social relations (Majumder & Tripathi, 2023). In this sense, they operate as cultural environments where values are negotiated, practices are transformed, and ways of life are shaped.

Analyzing their operation thus requires an ecological and relational approach—one that considers not only their technical aspects, but also their symbolic and cultural dimensions.

Cultural Dimensions of Technological Ecosystems

From this perspective, it becomes essential to incorporate the concept of *cultural impact*, understood as the set of transformative effects that a technological system or experience can generate for individuals and communities (Biedermann et al., 2024). To classify indicators of cultural impact, these authors propose three categories—individual, collective, and heritage—which provide a useful operational framework for analyzing the cultural dimension of digital ecosystems.

Impact on the Individual

This category refers to the effects that a system or experience may have at a personal level, encompassing indicators such as learning and its legacy; generated identity; symbolic and aesthetic values; emotions evoked by different experiences; and mechanisms of reflection and critical thinking (Biedermann et al., 2024). In technological ecosystems, this dimension may translate into changes in self-perception, emotional bonds with technology, and transformations in attention, thinking, or everyday behavior. Following the indicators proposed by Biedermann et al. (2024), technological ecosystems enable new expressive formats and distributed forms of cultural creation (creativity), promote hybrid identities shaped by digital interaction (identity), introduce new visual and sensory logics in technologically mediated experiences (aesthetics), and transform emotional connections through new experiences such as immersion or gamification (emotion). However, all these processes may either stimulate or inhibit critical reflection (reflection), depending on the design and intentionality of the interfaces—so the impact is not necessarily positive.

A sixth indicator can be added to this framework: everyday practices. Technological ecosystems influence the configuration of daily routines, reshaping how people organize their time, habits, and relationships with their environment. Technologies such as fitness trackers, virtual assistants, or smart home devices introduce automated dynamics into tasks like sleep management, climate control, or environmental interaction. Light, temperature, and sound can be automatically adjusted based on context, reducing direct intervention and altering the sensory relationship with space (Garg & Cui, 2022). Automated reminders and routine tracking can support independent living (Søraa et al., 2021), thereby redefining notions of autonomy and dependence. Seen in this light, technologies do not merely serve operational functions—they also shift decision-making toward automated systems that shape daily behaviors (Shanmugasundaram & Tamilarasu, 2023; Spinelli et al., 2020). Their deepest effect lies in the reconfiguration of habits, everyday decisions, and the perceptual frameworks through which experience is interpreted.

Collective Impact

This category focuses on symbolic cohesion among individuals, based on a sense of belonging to a community and its values, on the recognition of diverse groups and their expressions, and on the management of uncertainty in the face of future changes (Biedermann et al., 2024). In this sense, digital ecosystems can enable new forms of community, participation, and cooperation—for instance, through collective responses to crises—but they can also produce exclusion, polarization, or social fragmentation. Following the indicators proposed by Biedermann et al. (2024), technological ecosystems can facilitate the construction of symbolic communities and networks of affinity (belonging), enhance the visibility of minority expressions (diversity), and, at the same time, promote social bubbles or cultural tensions driven by the speed of change, misinformation, or the erosion of shared references (uncertainty).

Technological ecosystems also intervene in the construction of norms and values. By automating functions and distributing decision-making, they reshape concepts such as autonomy, privacy, and responsibility (Majumder & Tripathi, 2023; Verbeek, 2005). The technologies that constitute them embed social values in their design (Alsaleh, 2024), defining which behaviors are to be reinforced, monitored, or corrected—without requiring explicit consensus. For example, a home surveillance camera that activates and stores data by default normalizes surveillance as a form of care, shifting traditional boundaries of intimacy and prioritizing security over privacy. This transforms collective perceptions of what is acceptable (Selinger & Rhee, 2021). These logics extend across multiple domains such as health or education, where principles like efficiency or control are embedded as desirable values, reshaping technological coexistence at the community level (Grossi et al., 2025).

Beyond their role in shaping norms and values, technological ecosystems actively contribute to the formation of sociotechnical imaginaries—collective visions of possible futures organized around technological development (Jasanoff, 2015). These imaginaries are neither neutral nor homogeneous. They influence political decisions, innovation agendas, and criteria for social legitimacy (Grossi et al., 2025). Moreover, depending on context and cultural values, the same technology may represent progress and autonomy for some groups and surveillance or exclusion for others (Alsaleh, 2024; Mager & Katzenbach, 2021). The case of smart cities illustrates this tension: while their promoters frame them as efficient and sustainable environments, other social actors perceive them as technocratic or exclusionary spaces (Grossi et al., 2025). This plurality of interpretations reveals that technological ecosystems are not merely embedded within existing cultural frameworks—they also transform them. In this sense, imaginaries contribute to reorganizing collective symbolic references—what is possible, legitimate, or desirable—which guide decision-making about the future.

Impact on Heritage

This category relates to the production, transmission, and legitimization of memory and cultural knowledge. It includes aspects such as authenticity, heritage preservation, and the dissemination, development, and transfer of cultural and technological resources (Biedermann et al., 2024). Within this framework, technological ecosystems become key agents in the transformation of heritage, as they influence what is preserved, how it is represented, and which cultural narratives are transmitted or resignified. They thus act as new mediators of collective memory. Following the indicators proposed by Biedermann et al. (2024), technological ecosystems enable new forms of conservation and archiving (preservation) and can amplify their reach thanks to the accessibility provided by the transmedia circulation of content (dissemination); however, they also raise questions about the authenticity of works, their symbolic contextualization, and the protection of intellectual property (authenticity).

Technologies such as augmented reality, virtual reality, and digital archives support long-term heritage preservation, facilitate global access, and reinforce the intergenerational transmission of cultural identities (Xiaojuan, 2023). In addition, they generate immersive experiences that allow users to relive historical events, explore reconstructed cultural environments, or access previously unreachable works—expanding audiences, increasing the visibility of local expressions, and strengthening the connection with cultural memory, while also fostering diversity (Avcu & Koçoğlu, 2025).

However, these same processes involve decisions that are not neutral: what to preserve, how to display it, and which meanings to prioritize (Avcu & Koçoğlu, 2025). Without a critical approach, digitalization can trivialize or commodify heritage, eroding its symbolic value. In this sense, preserving heritage in digital contexts requires acknowledging the tensions between accessibility, representation, and authenticity—avoiding the simplification of complex content and advocating for a culturally situated and ethically grounded approach, in contrast to technocentric uses of these tools.

Culture as a Filter for Technological Acceptance

While the previous sections analyzed how technological ecosystems generate cultural transformations at individual, collective, and heritage levels, this section addresses the inverse process: how cultural frameworks condition the acceptance, adoption, and use of these technologies. The ways in which digital ecosystems are received, valued, or rejected are shaped by beliefs, norms, aesthetics, and prior experiences that define the symbolic framework of individuals and communities (Selwyn & Facer, 2007). In this sense, culture functions as a filter that decisively influences processes of technological appropriation. Among the main cultural factors that affect such acceptance, the following stand out:

Cultural values. The reception of a technology largely depends on its alignment with the values of the context in which it is introduced (Majumder & Tripathi, 2023). Concepts such as autonomy, privacy, hierarchy, or collectivism acquire specific meanings according to culture, affecting the perception of usefulness or intrusion (Majumder & Tripathi, 2023; Shin & Park, 2017). For example, a self-care app may be perceived as either useful or intrusive, depending on the value attributed to self-sufficiency versus shared caregiving.

Multiculturalism. In culturally diverse settings, the same technology may be interpreted and used in very different ways, depending on local expectations and social norms. A comparative study between Germany and Turkey on assistive technologies for older adults illustrates this contrast (Offermann-van Heek et al., 2021): whereas in Germany, individual autonomy and personal control over one's health were valued, in Turkey, family companionship and delegation to caregiving networks were prioritized. This difference directly influenced both the perceived functionality and the degree of acceptance of the same systems.

Previous experiences and relationships with technology. Trust or mistrust toward digital technologies is also shaped by prior experience. Technical failures, access gaps, or repeated exposure to techno-pessimistic narratives can create persistent symbolic barriers (Talukder et al., 2020; Ziefle & Rocker, 2010). In many cases, non-adoption does not result from voluntary rejection but rather from structural forms of socio-digital exclusion (Grošelj, 2021).

Aesthetics and stigma. The visual appearance and symbolic language of a device deeply influence its acceptance (Bardzell et al., 2018; Spooner et al., 2011). A lack of aesthetic or functional personalization may lead to rejection if the design conflicts with the user's identity (Goffman, 2023; Moller & Kettley, 2017). For instance, in stigmatized settings such as those involving dependency, medical aesthetics may be rejected due to their association with illness or old age (Spagnolli et al., 2014; Talukder et al., 2020).

Taken together, these factors demonstrate that technology cannot be designed according to a universalist logic. Digital systems must be conceived as situated mediations, whose acceptance depends on their capacity to resonate with the symbolic and sociocultural frameworks of the communities they aim to serve (Boger, 2022; Shin & Park, 2017).

Integration Processes of Technological Ecosystems

The initial acceptance of a technology does not guarantee its sustained integration into everyday life. In order to become an active part of socio-cultural environments, technological ecosystems must undergo progressive processes of adaptation, resignification, and symbolic continuity. These processes, grounded in situated appropriation by users, form the cultural basis of their long-term sustainability (Shin & Park, 2017).

One useful concept for understanding this integration is **technological domestication** (Silverstone & Haddon, 1996), which conceives adoption as a gradual trajectory in which technology moves from novelty to familiarity. This process involves practical dimensions (placement, configuration), cognitive dimensions (understanding), symbolic dimensions (affective bonds), and social dimensions (roles within the environment). Sustainability does not depend solely on technical efficiency, but on the system's ability to embed itself in existing ways of life and generate meaning (Søraa et al., 2021). Designing from this perspective requires not only optimizing functionality, but also enabling processes of cultural reception. It calls for sensitivity to everyday rhythms, symbolic references, and the values of the context in which the technology is introduced.

As technologies become embedded in everyday life, their uses may diverge from their original purposes. Users **reinterpret and resignify** technological ecosystems according to their cultural, emotional, and practical frameworks, generating new, emergent meanings (Grošelj, 2021; Søraa et al., 2021). Far from being deviations, these transformations reflect legitimate forms of cultural appropriation. Such changes may be expressed through the repurposing of certain functions, the symbolic use of technical components, aesthetic or identity-based adaptations (Grošelj, 2021), or the reconfiguration of social relationships through new uses and modes of connection (Blanco et al., 2019). Along these lines, Grošelj (2021) introduces the concept of **re-domestication** to describe the transformation in the role of already-adopted technologies when new devices are introduced or living conditions change. These resignifications respond not only to technical evolution, but also to symbolic adaptations that allow technologies to maintain their relevance in changing environments.

For a technological ecosystem to endure, its sustainability must be considered across multiple dimensions. In particular, **cultural sustainability** depends on its ability to retain meaning over time and evolve alongside the people and contexts it inhabits (Avcu & Koçoğlu, 2025). This symbolic continuity is key to avoiding abandonment, obsolescence, or rejection. Thinking in terms of cultural sustainability involves designing systems that are open, flexible, and adaptable, capable of engaging with different ways of life.

Challenges for the Cultural Sustainability of Technological Ecosystems

Designing culturally sustainable technological ecosystems is a critical challenge in a context of accelerated digital transformation. Factors such as cultural standardization, symbolic decontextualization, excessive automation, access asymmetries, or the absence of maintenance strategies threaten their consolidation and widen adoption gaps.

Many technological ecosystems—especially those developed by large platforms—adhere to standardized models designed for global scalability. This logic promotes efficiency but is often based on the values and functional structures of dominant cultures, which may not align with local contexts. The result is cultural **homogenization** (Avcu & Koçoğlu, 2025; Shin & Park, 2017). As cultural specificity is diluted, epistemic hierarchies are reproduced, excluding alternative ways of imagining the future (Alsaleh, 2024).

When cultural practices and contexts are digitized, they do not always retain their original meaning. Their **decontextualization** or resignification may transform or fragment the value of practices, objects, or relationships; impoverish expressions such as community rituals or oral narratives; or reduce them to depersonalized content or consumer products. Although digitalization allows for greater global visibility, it also entails decisions that affect cultural representation, legitimacy, and authenticity (Avcu & Koçoğlu, 2025; Noble, 2018).

Automation, while contributing to efficiency, can undermine user agency when implemented without room for interpretation or intervention. Under predictive and control-oriented logics, systems tend to standardize behavior, weakening relational and symbolic capacities linked to identity. This can result in a **passive technological relationship** in which users adapt their practices to system automatisms, without space to question or reconfigure them (Stewart et al., 2024).

Finally, unequal access to technological infrastructures creates exclusion, limiting the participation of certain groups in digital environments and affecting the transmission of their cultural heritage (Noble, 2018). This **sociotechnical divide** is not only the result of material scarcity, but also the result of the limited involvement of diverse actors in the design and configuration of systems (Avcu & Koçoğlu, 2025; Blanco et al., 2019). Designed from urban and technocratic logics, many ecosystems overlook the conditions of rural or peripheral areas, or of communities at risk of exclusion—as well as the symbolic expressions of their users, which are often excluded from design criteria (Alsaleh, 2024; Noble, 2018).

Design Strategies for Culturally Sensitive Digital Ecosystems

As argued throughout this text, the sustainability of technological ecosystems does not depend solely on technical performance. Their long-term viability requires meaningful integration into the cultural contexts in which they operate. Preserving this cultural dimension involves establishing ethical, legal, and strategic frameworks that promote equity and diversity as structural principles of the digital ecosystem.

Moreover, recognizing the user as an active agent is a key principle in the design of culturally sustainable digital ecosystems. This capacity for action—or ***agency***—is essential for sustaining meaningful relationships between people and digital systems. Designing with this premise means not only facilitating participation, but also creating open structures that allow for diverse forms of appropriation, respect local autonomy, and foster critical literacy of the digital environment (Avcu & Koçoğlu, 2025). This approach moves beyond seeing the user as an isolated individual, toward understanding them as part of interdependent communities. The design of digital ecosystems must respond to this shift by enabling convergence between physical and digital worlds as parallel environments that support both personal growth and the collective construction of meaning (Farfieva, 2024). Instead of designing for the "individual," the goal is to design for the individual ***and*** their broader community, recognizing relationality as a structural dimension of any digital system (Blanco et al., 2026a).

Culturally sensitive design not only requires adaptation to existing contexts, but also the creation of spaces for new forms of relationship to emerge between people, technology, and their environment. Design can act as a **symbolic mediator**, enabling meaningful connections between digital systems and evolving cultural frameworks. From this perspective, traditional design logics must be reconfigured. Beyond prioritizing efficiency or scalability, it is essential to integrate functional, cultural, relational, ethical, and experiential criteria from the outset (Avcu & Koçoğlu, 2025; Rosenthal et al., 2021). As Li and Zhang (2023) state, a more sustainable and culturally aware approach to product design contributes to reinforcing cultural identity and promoting long-term social and environmental value.

In this context, approaches such as human-centered design, co-design, inclusive design, or iterative methodologies provide useful frameworks for incorporating diversity, active participation, and contextual adaptability into the development of digital ecosystems (Boger, 2022; Spinelli et al., 2020; Rinaldi & Kianfar, 2021). However, implementing these approaches effectively in complex technological systems requires multidisciplinary teams and processes capable of integrating technical, social, cultural, and human knowledge (Blanco, 2016; Blanco et al., 2017; Blanco et al., 2025; García-Peñalvo, 2013; Rosenthal et al., 2021).

In recent years, methodologies aimed at designing digital ecosystems have begun to emerge in fields such as the Internet of Things, artificial intelligence, and complex digital services. Alongside performance-focused approaches, several proposals have taken shape that explicitly incorporate cross-cutting dimensions, with the goal of developing systems that not only function but are also inhabitable, meaningful, and sustainable in diverse contexts. These methodologies address various levels of the design process. Some incorporate cultural principles explicitly, while others focus on user-centered design, participatory methods, or contextual adaptability. This diversity reflects the multiple ways sociocultural sustainability can be addressed through design. All of these approaches incorporate multidisciplinary perspectives to a greater or lesser extent. Table 7.1 presents a representative selection of these methodologies, indicating their relationship to the challenges identified in previous sections.

Based on this compilation, it is possible to identify different methodological orientations that address the challenges of designing digital ecosystems from complementary perspectives.

Culturally Conscious and Participatory Approaches: Methodologies such as Value Sensitive Citizen Science (VSCS) framework (Mahajan & Helbing, 2025), Community (Blanco, Marco et al., 2026a), or the Culturally Localized User Experience (CLUE) framework (Sun, 2012) place cultural diversity, local imaginaries, and/or community interrelations at the center of the design process. These approaches recognize the plurality of knowledge systems and languages and, in addition to

Table 7.1 Selection of methodologies for the design of technological ecosystems addressing (1) cultural homogenization, (2) sociotechnical exclusion, (3) symbolic decontextualization, and (4) loss of agency

Methodology	*Type of methodology*	*Cultural approach*	*Challenges addressed*
CLUE Theoretical Framework (Sun, 2012)	Intercultural theoretical framework focused on situated UX. Intercultural design centered on situated experience, mediation, and local adaptation	Cultural and localized approach	1, 3
Community (Blanco, Siguín et al., 2026b)	Multidisciplinary methodology for the design of digital ecosystems based on relational needs and interaction spaces. Situated and participatory design	High localization, diversity, situated evaluation based on symbolic resonance with relational needs	1, 2, 3, 4
VSCS (Mahajan & Helbing, 2025)	Hybrid framework: VSD + citizen science + scenario-based logic. Translation of cultural values into technical requirements with structured participation	Explicit (local values, epistemic justice)	1, 2, 3, 4
PawEn (Fauquex et al., 2015)	Structured design in 7 iterative phases combining design thinking and UCD for IoT environments	Partial (contextual and user-centered, without explicit cultural dimension)	2, 4
DSTS (Winby & Mohrman, 2018)	Sociotechnical digital design applied to ecosystems. Iterative design of multi-actor technological ecosystems	Implicit (user-centered and co-creation focused)	2, 3, 4
Cosica (Blanco, Marco et al., 2026a)	Structured multidisciplinary methodology. Visual and matrix-based framework for coordinating design, technology, and technological literacy in IoT design processes	Partial (collaborative and open, without explicit cultural dimension)	2, 4
CCI-DBE Model (Hsu et al., 2023)	Strategic framework based on RCA for co-creation in CCIs. Ecosystemic sectoral design with integration of stakeholders and cultural experience	Explicit (symbolic and cultural value in ecosystems)	1, 3

improving the user experience, incorporate specific tools to translate these values into technical requirements. They promote epistemic justice (the recognition and equitable inclusion of knowledge systems, languages, and worldviews from diverse communities), symbolic resonance, and user agency in complex contexts—thus addressing multiple challenges related to sociocultural sustainability.

Hybrid Approaches Applied to IoT Ecosystems: Frameworks such as Cosica (Blanco, Siguín et al., 2026b), PawEn People-Aware Environment (PawEn) (Fauquex et al., 2015), and DSTS Digital Sociotechnical System Design (DSTS) (Winby & Mohrman, 2018) integrate structured technical tools with a degree of contextual awareness. Although their main emphasis often lies in defining technical specifications, ensuring interoperability, or optimizing scalability, they also incorporate collaborative, iterative, and multidisciplinary practices. In these approaches, multidisciplinarity is not merely an operational choice but a structural requirement that enables the navigation of both the vertical and horizontal complexities inherent in digital ecosystems.

Sector-Specific Frameworks for Cultural and Creative Industries: The CCI-DBE Cultural and Creative Industries Digital Business Ecosystem (CCI-DBE) model (Hsu et al., 2023) exemplifies a more strategic, sector-specific approach tailored to cultural and creative industries. Although it does not constitute a fully articulated design methodology, it provides valuable insights for embedding cultural and symbolic experience into co-creation processes with multiple stakeholders. This framework expands the boundaries of technological design by linking it to cultural policy-making, participatory governance, and long-term sustainability goals.

This classification is neither exhaustive nor hierarchical; rather, it offers an overview of how different methodologies incorporate elements such as participation, cultural diversity, adaptability, and user agency in relation to the design context.

Conclusions and Future Research

This chapter has proposed a cultural perspective on technological ecosystems, understood as interconnected sociotechnical systems in which devices, users, values, practices, and contexts converge. From this perspective, three fundamental axes have been explored: (1) the evolution of the product as a cultural node within networks of meaning; (2) the configuration of digital ecosystems as living environments that shape relationships, ways of life, and power dynamics; and (3) the need for culturally sensitive design approaches capable of sustaining these systems.

In this scenario, user agency, contextual adaptability, and multidisciplinarity emerge as fundamental principles for designing systems that are efficient, inhabitable, meaningful, and sustainable. The methodologies reviewed demonstrate significant progress, yet they also raise important questions for future research aimed at achieving a systemic understanding of the cultural sustainability of digital ecosystems. Among others: Can we measure the cultural impact that a technological system has on a community—or vice versa? Can we anticipate how such systems will evolve over time? Which indicators, methodologies, or epistemological frameworks are most appropriate for evaluating the cultural sustainability of technological interventions? And which actors should be considered in the multidisciplinary and systemic design of technological ecosystems? As Majumder and Tripathi (2023) state, in an increasingly interconnected world where the digital permeates every aspect of daily life, the act of designing technological tools and systems becomes inseparable from designing ways of being.

References

Alsaleh, A. (2024). The impact of technological advancement on culture and society. *Scientific Reports, 14*(1), 32140.

Anwar, M. J., & Gill, A. Q. (2019, July). A review of the seven modelling approaches for digital ecosystem architecture. In *2019 IEEE 21st conference on business informatics (CBI)* (Vol. 1, pp. 94–103). IEEE.

Avcu, K. M., & Koçoğlu, E. (2025). Preservation of cultural identity in digital ecosystems. *Education Quarterly Reviews, 8*(1). https://doi.org/10.31014/aior.1993.08.01.552

Bardzell, J., Bardzell, S., & Liu, S. Y. (2018). Beautifying IoT: The internet of things as a cultural agenda. In *Social internet of things* (pp. 3–21). Springer International Publishing.

Biedermann, A. M., Muñoz López, N., Santolaya Sáenz, J. L., Asión-Suñer, L., & Galán Pérez, F. J. (2024). Methodological framework for integrating cultural impact in sustainability assessments of cultural events. *Sustainability, 16*(16), 6893.

Blanco, T. (2016). *Design methodologies as a platform for cross-disciplinarity in technological projects: Surfing disciplines* [Doctoral dissertation]. Universitat Politècnica de València. https://doi.org/10.4995/Thesis/10251/77149

Blanco, T., Casas, R., Manchado-Pérez, E., Asensio, Á., & López-Pérez, J. M. (2017). From the islands of knowledge to a shared understanding: interdisciplinarity and technology literacy for innovation in smart electronic product design. *International Journal of Technology and Design Education, 27*(2), 329–362.

Blanco, T., Marco, A., Casas, R. (2026a). *Personas Are Running the Community: Why Tech Keep Driving Us Crazy in Interrelational Scenarios and a Contribution to Sanity.* Preprint.

Blanco, T., Siguín, M., Marco, A., Marín, J., Casas, R. (2026b), Cosica: A multidisciplinary methodology for designing thing–cloud–app systems in the Internet of Things. Preprint.

Blanco, T., Casas, R., Marco, A., & Martínez, I. (2019). Micro ad-hoc health social networks (uHSN). Design and evaluation of a social-based solution for patient support. *Journal of Biomedical Informatics, 89*, 68–80.

Blanco-Novoa, Ó., Fraga-Lamas, P., A. Vilar-Montesinos, M., & Fernández-Caramés, T. M. (2020). Creating the internet of augmented things: An open-source framework to make IoT devices and augmented and mixed reality systems talk to each other. *Sensors, 20*(11), 3328.

Boger, J. (2022, June). Culture change, human-centered design, and ethical by design as transactional cornerstone concepts in the development of technology for supporting aging. In Proceedings of the 15th International Conference on PErvasive Technologies Related to Assistive Environments (pp. 556–561).

Bresciani, S., Ferraris, A., Romano, M., & Santoro, G. (2021). Digital ecosystems. In *Digital transformation Management for Agile Organizations: A compass to sail the digital world* (pp. 153–165). Emerald Publishing Limited.

Briscoe, G. (2009). Digital ecosystems. [Doctoral dissertation]. arXiv:0909.3423.

Carvalho, L., & Goodyear, P. (2018). Design, learning networks and service innovation. *Design Studies, 55*, 27–53.

Cash, P., Holm-Hansen, C., Borum, S., Christensen, M. L., & Mai, Y. (2017). Uniting individual and collective concerns through design: Priming across the senses. *Design Studies, 49*, 32–65.

Dulsrud, A., & Bygstad, B. (2022). Digital ecosystems as social institutions: Exploring the role of consumption through four research streams of digital ecosystems. *Consumption and Society, 1*(1), 99–119.

Farfieva, K. A. (2024). The virtual world and personal development: The impact of digital environments on individual growth. *Excellencia: International multi-disciplinary Journal of Education, 2*(10), 547–561.

Fauquex, M., Goyal, S., Evequoz, F., & Bocchi, Y. (2015, December). Creating people-aware IoT applications by combining design thinking and user-centered design methods. In *2015 IEEE 2nd World Forum on Internet of Things (WF-IoT)* (pp. 57–62). IEEE.

García-Holgado, A., & García-Peñalvo, F. J. (2018, May). Human interaction in learning ecosystems based on open source solutions. In *International conference on learning and collaboration technologies* (pp. 218–232). Springer International Publishing.

García-Peñalvo, F. J. (2013). *Multiculturalism in technology-based education: Case studies on ICT-supported*. Information Science Reference.

Garg, R., & Cui, H. (2022). Social contexts, agency, and conflicts: Exploring critical aspects of design for future smart home technologies. *ACM Transactions on Computer-Human Interaction, 29*(2), 1–30.

Goffman, E. (2023). The presentation of self in everyday life. In C. Lemert (Ed.), *Social theory re-wired* (pp. 450–459). Routledge.

Grošelj, D. (2021). Re-domestication of internet technologies: Digital exclusion or digital choice? *Journal of Computer-Mediated Communication, 26*(6), 422–440.

Grossi, G., Dobija, D., Staniszewska, Z., & Kaczmarek-Ciesielska, D. (2025). Exploring socio-technical imaginaries on smart city sustainability. *Sustainability Accounting, Management and Policy Journal, 16*(7), 186–217.

Hsu, S. F., Wans, X. Y., Yang, C. H., & Liou, S. (2023). Crafting digital business ecosystems for cultural and creative industries. *International Journal of Research in Business and Social Science, 12*(9), 1–15.

Jasanoff, S. (2015). Future imperfect: Science, technology, and the imaginations of modernity. In *Dreamscapes of modernity: Sociotechnical imaginaries and the fabrication of power* (pp. 1–33). University of Chicago Press.

Jordan, P. W. (2000). *Designing pleasurable products: An introduction to the new human factors*. CRC Press.

Klein, J. T. (1994). *Notes toward a social epistemology of Transdisciplinarity*. Communication au Premier Congrès Mondial de la Transdisciplinarité.

Li, X., & Zhang, Y. (2023). Cultural product design: Principles and technique for sustainable and innovative product development. In *Advances in industrial design* (pp. 123–134). Springer. https://doi.org/10.1007/978-3-031-63061-3_15

Mager, A., & Katzenbach, C. (2021). Future imaginaries in the making and governing of digital technology: Multiple, contested, commodified. *New Media & Society, 23*(2), 223–236.

Mahajan, S., & Helbing, D. (2025). Co-designing AI systems with value-sensitive citizen science. Available at SSRN 5217120.

Majumder, M., & Tripathi, A. K. (2023). Transformative power of technologies: Cultural transfer and globalization. *AI & SOCIETY, 38*(6), 2295–2303.

Manikas, K., & Hansen, K. M. (2013). Software ecosystems–a systematic literature review. *Journal of Systems and Software, 86*(5), 1294–1306.

Moller, T., & Kettley, S. (2017). Wearable health technology design: A humanist accessory approach. *International Journal of Design, 11*(3), 1–49.

Noble, S. U. (2018). *Algorithms of oppression: How search engines reinforce racism*. NYU Press.

Norman, D. A., & Verganti, R. (2014). Incremental and radical innovation: Design research vs. technology and meaning change. *Design Issues, 30*(1), 78–96.

Offermann-van Heek, J., Wilkowska, W., & Ziefle, M. (2021). Cultural impact on perceptions of aging, care, and lifelogging technology: A comparison between

Turkey and Germany. International Journal of Human–Computer Interaction, 37(2), 156–168.

Omol, E. J. (2024). Organizational digital transformation: From evolution to future trends. *Digital Transformation and Society, 3*(3), 240–256.

Perera, C., Zaslavsky, A., Christen, P., & Georgakopoulos, D. (2014). Context aware computing for the internet of things: A survey. *Communications Surveys & Tutorials, IEEE, 16*(1), 414–454.

Petrova, M., & Marinova, K. (2021). Digital ecosystem: Nature, types & opportunities for value creation. In *International scientific conference on Innovations in Digital Economy*, pp. 71–85.

Rinaldi, A., & Kianfar, K. (2021, May). Digital technologies as opportunity for facilitating social inclusion and multicultural dialogue. In *Congress of the international ergonomics association* (pp. 325–333). Springer International Publishing.

Rosenthal, A., Guedes, A. M. M., dos Santos, K. M. O., & Deliza, R. (2021). Healthy food innovation in sustainable food system 4.0: Integration of entrepreneurship, research, and education. *Current Opinion in Food Science, 42*, 215–223.

Santhi, A. R., & Muthuswamy, P. (2023). Industry 5.0 or industry 4.0 S? Introduction to industry 4.0 and a peek into the prospective industry 5.0 technologies. *International Journal on Interactive Design and Manufacturing (IJIDeM), 17*(2), 947–979.

Selinger, E., & Rhee, H. J. (2021). Normalizing surveillance. *Sats, 22*(1), 49–74.

Selwyn, N., & Facer, K. (2007). Beyond the digital divide: Rethinking digital inclusion for the 21st century. Bristol, England. https://doi.org/10.1002/ncr.

Shanmugasundaram, M., & Tamilarasu, A. (2023). The impact of digital technology, social media, and artificial intelligence on cognitive functions: A review. *Frontiers in Cognition, 2*, 1203077.

Shin, D. H., & Jin Park, Y. (2017). Understanding the Internet of Things ecosystem: Multi-level analysis of users, society, and ecology. *Digital Policy, Regulation and Governance, 19*(1), 77–100.

Silverstone, R., & Haddon, L. (1996). Design and the domestication of information and communication technologies: Technical change and everyday life. In R. Silverstone & R. Mansell (Eds.), *Communicating by design: The politics of information and communication technologies* (pp. 44–74). Oxford University Press.

Søraa, R. A., Nyvoll, P., Tøndel, G., Fosch-Villaronga, E., & Serrano, J. A. (2021). The social dimension of domesticating technology: Interactions between older adults, caregivers, and robots in the home. *Technological Forecasting and Social Change, 167*, 120678.

Spagnolli, A., Guardigli, E., Orso, V., Varotto, A., & Gamberini, L. (2014). Measuring User Acceptance of Wearable Symbiotic Devices: Validation Study Across Application Scenarios. In Proceedings (pp. 87–98). https://doi.org/10.1007/978-3-319-13500-7_7

Spinelli, G., Micocci, M., Tsekleves, E., Wang, Y. H., Martin, W., Lim, Y., & Shamim, U. (2020). Contemporary themes in the design of AT for the ageing population: Materiality, co-design and cultural influences. In *Design of Assistive Technology for ageing populations* (pp. 41–58). Springer.

Spooner, D., Raynauld, J., & Lalande, P. (2011) Proposed framework for transdisciplinary product and process design education. Proceedings of the 7th international CDIO conference, Technical University of Denmark.

Stewart, K., Perren, R., Chambers, C., & Zulauf, R. (2024). In tech we rely: How technology dependence fuels consumer vulnerability. *Journal of Consumer Affairs, 58*(4), 905–945.

Streitz, N. (2015). Themes & Issues - Smart Future Initiative. Smart-future.net. Retrieved November 7, 2015, from http://www.smart-future.net/3.html

Sun, H. (2012). *Cross-cultural technology design: Creating culture-sensitive technology for local users.* OUP USA.

Sussan, F., & Acs, Z. J. (2017). The digital entrepreneurial ecosystem. *Small Business Economics, 49*(1), 55–73.

Talukder, M. S., Sorwar, G., Bao, Y., Ahmed, J. U., & Palash, M. A. S. (2020). Predicting antecedents of wearable healthcare technology acceptance by elderly: A combined SEM-neural network approach. *Technological Forecasting and Social Change, 150*, 119793.

Tsujimoto, M., Kajikawa, Y., Tomita, J., & Matsumoto, Y. (2018). A review of the ecosystem concept—Towards coherent ecosystem design. *Technological Forecasting and Social Change, 136*, 49–58.

Ulaga, W., Kondis, A., & McTeague, L. (2013). From product to service: Navigating the transition. Insight IMD, (26), 1–4.

Vargo, S. L., Fehrer, J. A., Wieland, H., & Nariswari, A. (2024). The nature and fundamental elements of digital service innovation. *Journal of Service Management, 35*(2), 227–252.

Verbeek, P. P. (2005). *What things do: Philosophical reflections on technology, agency, and design.* Penn State Press.

Vroom, R. W., & Horvath, I. (2014). Editorial: How are product development and engineering processes enhanced by involving research. *Journal of Design Research, 12*(1/2), 2–9.

Winby, S., & Mohrman, S. A. (2018). Digital sociotechnical system design. *The Journal of Applied Behavioral Science, 54*(4), 399–423.

Xiaojuan, J. (2023). Technology and culture in the digital era. *Social Sciences in China, 44*(1), 4–24.

Ziefle, M., & Röcker, C. (2010, March). Acceptance of pervasive healthcare systems: A comparison of different implementation concepts. In *2010 4th International Conference on Pervasive Computing Technologies for Healthcare* (pp. 1–6). IEEE.

CHAPTER 8

New Methodologies for Assessing the Impact of Cultural Projects in Rural Areas: The Preliminary Design Phase

Rosana Sanz-Segura, Carlos Romero-Piqueras, Giordana Gallina, and Eduardo Manchado-Perez

Introduction

The Role of Culture in Rural Development: Territorial Imbalances and Policy Blind Spots

In the workbook *Intuiciones sobre la Cultura y la Ruralidad* (*Insights on Culture and Rurality*), Alfons Martinell Sempere states in his foreword, "In order to move towards a greater presence of the cultural dimension in

R. Sanz-Segura (✉)
Department of Design and Manufacturing Engineering, University of Zaragoza, Zaragoza, Spain

OAAEP Group, Funded by the Government of Aragón with FEDER Funds, Zaragoza, Spain
e-mail: rsanz@unizar.es

A. M. Biedermann, F. J. Galán Pérez (eds.), *Interconnections between Culture and Sustainability*,
https://doi.org/10.1007/978-3-032-21061-6_8

the 2030 Agenda, it is also necessary to have a closer approach to local and territorial realities, their particularities and contact with their actors". Although the majority of the world's population lives in urban areas, the perspective of rural communities remains essential, not only to guarantee the cultural rights of their citizenship but also to preserve the tangible and intangible heritage that embodies cultural diversity (Martinell et al., 2023).

The concentration of population and activities in an increasingly small number of urban areas exacerbates the inequalities in access to culture between urban and rural contexts and highlights the need for approaches that integrate not only the creative dimension but also the methodological and sustainable principles in the design of cultural initiatives. Furthermore, the rural context, which is often underrepresented in the scientific literature on creative industries, requires prioritized attention, if we aim to address the issues concerning the risk of socio-territorial imbalance faced by Spain's territorial model, and a large part of Europe (Ministerio para la Transición Ecológica y el Reto Demográfico, *Ministry for the Ecological Transition and the Demographic Challenge*, 2025a, 2025b).

We advocate that the contribution of cultural industries in rural areas must be realized with the same emphasis and rigor as in urban areas, thereby reducing access disparities and fostering social and economic growth opportunities for local communities. Nevertheless, and as Martinell points out, although the 2030 Agenda includes the rural environment in the goal 11.a: "Support positive economic, social and environmental links between urban, peri-urban and rural areas by strengthening national and regional development planning", it does not include the consideration that local and cultural life can constitute important development dynamics in rural areas. Moreover, it doesn't explicitly recognize culture and local life as potential drivers of rural transformation.

Similarly, although the EU Rural Observatory promotes knowledge generation and aims to enhance the collection and dissemination of data across the economic, social and environmental dimensions of rural areas (European Commission, 2025), it still excludes culture as a key vector for cohesion and territorial equity.

C. Romero-Piqueras • G. Gallina • E. Manchado-Perez
Department of Design and Manufacturing Engineering, University of Zaragoza, Zaragoza, Spain
e-mail: cromerop@unizar.es; ggallina@unizar.es; manchado@unizar.es

Territorial cohesion, environmental sustainability, and human well-being must be prioritized within a new paradigm aligned with the international agendas for sustainable development. In this regard, the World Conference on Cultural Policies and Sustainable Development, held in Spain in the second half of 2025 (Ministerio de Cultura, *Ministry of Culture*, 2025), has contributed to the momentum for defining a global cultural agenda, with the objective of establishing a specific Sustainable Development Goal (SDG) for culture. Thus, the cross-cutting and intersectional integration of culture across all SDGs is a key factor to achieve more impactful and equitable sustainable development outcomes.

Culture and Sustainability: A Cross-Cutting Perspective

Within this framework, sustainability is conceived as a guiding principle that goes beyond its environmental dimension to also encompass cultural continuity, social cohesion, and the economic viability of events and projects carried out in low-density territories. The integration of sustainability into emerging cultural policies has opened up the debate on the role of rural areas as drivers of cultural decentralization.

Recent reports, such as those by the Ministry of Social Rights and the 2030 Agenda, and the Spanish Network for Sustainable Development (Ecológica y el Reto Demográfico, *Ministry for the Ecological Transition and the Demographic Challenge*), stress the urgency of making the Sustainable Development Goals (SDGs) accessible and meaningful to diverse audiences. This involves advancing cultural and socio-economic rights while equipping the cultural sector with concrete tools for evaluation, funding, training, and the dissemination of methodological best practices.

Moreover, these reports highlight the strategic importance of rural areas as essential arenas for strengthening cultural democracy and fostering citizen participation, both of which are vital levers in addressing the challenges of depopulation (Fig. 8.1).

Toward a Methodological Framework for Cultural Design

In response to these challenges, this chapter presents a human-centered design framework focused on the early stages of cultural project development in rural areas. This methodology, developed by the Observatorio de Creación Cultural en Entornos Rurales de Aragón (*Aragon Observatory*

Fig. 8.1 Mozota (Zaragoza, Spain). Everyday landscape and rural scenery of a small municipality. (Source: Pinterest (Cultura Inquieta))

for Cultural Creation in Rural Environments) or OCREA, prioritizes participatory processes, accessibility, and responsiveness to local needs. The proposed approach seeks to establish a practical, replicable, and people-focused methodology that places community well-being and satisfaction at its core. It embraces participatory design as both a planning and evaluative tool, embedding mechanisms for continuous feedback and collaborative reflection throughout the lifecycle of cultural projects. Crucially, evaluation is not conceived as an external or isolated process, but as one co-constructed with the local population. By incorporating open channels for feedback at different stages, the methodology ensures that communities have an active role in defining success criteria, assessing outcomes, and adapting the project as needed. These participatory evaluation practices strengthen the legitimacy of the results and ensure that projects remain rooted in the lived experiences, aspirations, and changing priorities of the communities they aim to serve (Fig. 8.2).

Research Background and Methodological Foundations

This chapter builds upon prior work in cultural impact evaluation, particularly focusing on the role of four key stakeholder groups—citizens, public institutions, cultural professionals, and academia—in shaping effective and inclusive evaluation models. It draws specifically from two foundational publications: *Claves e Impactos de la Cultura en el Medio Rural* (*Keys and Impacts of Culture in Rural Areas*) (Sáez Pérez et al., 2024) and the *Nueva guía para la evaluación de las políticas culturales locales* (*New Guide*

Fig. 8.2 Engagement of local youth in the design and production of the "El Bosque Sonoro" music festival in Mozota (Zaragoza, Spain). (Source: OCRE)

for the Evaluation of Local Cultural Policies) (FEMP, 2022), with particular emphasis on indicators aligned with sustainability.

While the first reference explores the transformative potential of cultural activities for rural territories, the second offers a detailed system of indicators focusing on sustainability, gender equity, accessibility, and territorial integration. As paraphrased from Sáez Pérez et al. (2024): "The goal is for all these reflections and studies to generate strategies and empower the people living in these small communities, so that, individually and collectively, they can develop better, fuller and more active lives".

The human-centered design approach promoted here reflects a commitment to economic, social, and environmental sustainability, explicitly positioning the cultural dimension as a key contributor to our work. It is argued that genuine interconnections between culture and sustainability can only be achieved when creative practices strengthen local identity, reinforce social cohesion, and enhance community resilience, thereby ensuring cultural continuity and relevance within the rural context.

This experience is embedded in the evaluation model developed by Sáez Pérez et al. (2024), based on 2500 surveys and multiple interviews across various stakeholder groups. This model reinforces the use of longitudinal observation and context-sensitive diagnostics as tools for empowerment and co-construction of value. It also underscores that the role of promoters is not limited to implementation but extends to documentation and evaluation, enabling feedback loops that inform future cultural policy and community planning.

Cultural Project Design as Service Design

The Observatorio de Creación Cultural en Entornos Rurales de Aragón (*Aragon Observatory for Cultural Creation in Rural Environments*) (OCREA) is the Research & Development & Innovation project carried out by the authors of this chapter in collaboration with the non-profit social association OCRE. Its goal is to develop new methodological approaches to the design, management, and promotion of cultural projects adapted to the specific challenges and opportunities of rural territories.

The advancement and evolution of contemporary cultural practices requires the integration of the different dimensions of sustainability, alongside a systemic and strategic vision. This vision must also generate exogenous processes across other fields such as tourism, governance, law, social and academic research.

To this goal, OCRE applies its own methodology, which incorporates principles from open innovation and human-centered design methods and tools, and encompasses all stages of the design process:

1. Planning
2. Territorial studies and analysis
3. Cross-sectoral integration through the quadruple helix[1]
4. Conceptualization
5. Funding strategies
6. Technical and artistic production
7. Communication
8. Implementation and execution
9. Impact assessment
10. Dissemination of results

[1] Collaborative innovation model integrates four key groups: academia, industry, public institutions, and civil society. Its objective is to foster innovation through ongoing interactions among these stakeholders, promoting inclusive and effective solutions.

In collaboration with the School of Engineering and Architecture (EINA) at the University of Zaragoza, through its Research Transfer Office (OTRI), OCRE promotes the development of a multidisciplinary guide for designing cultural projects in rural areas, with a specific focus on the preliminary design phase. This early stage is understood as a critical moment for ensuring the viability and sustainability of cultural initiatives by aligning them with the territorial context, available resources, and the interests of local stakeholders (Fig. 8.3).

The preliminary design phase, also referred to as the conceptual or exploratory stage, is widely acknowledged in strategic and user-centered design literature. Authors such as Nigel Cross (2001) and Tim Brown (2008) highlight its value for uncovering real user needs and rapidly prototyping solutions. From a more technical standpoint, Ulrich and Eppinger (2016) identify this phase as key to ensuring feasibility and long-term project viability. In the field of service design, Stickdorn and Schneider (2012), as well as companies as Ideo (Brown & Wyatt, 2010), emphasize the importance of early-stage user engagement through participatory methods and agile prototyping.

The originality of our approach lies in adapting these principles to the specific context of rural cultural projects, integrating participatory design tools and tailored impact evaluation metrics. As Radice (2014) points out, the preliminary phase involves combining the motivational profiles of target audiences with institutional strategies and participatory practices. Its outcome is the project brief, which includes the refined main message, affective and cognitive goals, the storyline, and the take-home messages.

Fig. 8.3 Cultural mediation activities (2024) during the preliminary project phase of the OCREA Observatory, developed through the quadruple helix approach. (Source: OCRE)

Thus, the preliminary design phase constitutes the foundational study aimed at describing and structuring the essential elements required for the subsequent drawing up and development of a project. Designing without a method is not viable; it must follow a clearly defined and purpose-driven approach. This phase demands the acquisition of in-depth knowledge about the context in which the project will be implemented, the mastery of the technical conditions necessary for its execution, and the identification of all relevant stakeholders. Only by aligning these dimensions within a coherent and coordinated framework can the project ensure its viability and long-term impact.

Crucially, this early stage must also include the design of a context-sensitive evaluation strategy, particularly in rural territories, where traditional statistical data are often insufficient, fragmented, or entirely absent. The absence of reliable data in these settings reinforces the need for tailored methodologies that can effectively capture the complexity and specificity of rural cultural ecosystems.

Accordingly, the preliminary phase should define and structure the key components that will guide the project's development. It is not simply a preparatory exercise, but a strategic moment to integrate technical, territorial, and social dimensions into a cohesive plan. This includes the selection of evaluation tools and indicators capable of capturing both tangible and intangible impacts in a given locality.

The study *Claves e Impactos de la Cultura en el Medio Rural* (*Keys and Impacts of Culture in Rural Areas*) (Sáez Pérez et al., 2024) identifies this limitation and addresses it by proposing a triangulated evaluation methodology that combines direct observation, structured surveys, and interviews with diverse stakeholders. This approach also takes into account demographic variability and the scale of intervention, ensuring that the analysis remains sensitive to local dynamics and differentiated rural realities.

Existing frameworks for evaluating cultural impact frequently neglect the particular dynamics and specificities of rural contexts. Analyzing the outcomes of cultural projects in such settings enables the transfer of empirical experience into structured methodologies. As noted by Sáez Pérez et al. (2024), the most effective strategy for assessing impact involves examining its potential demographic, economic, reputational, and well-being-related dimensions (at both individual and community levels) through a triangulation of perspectives from promoting agents, expert stakeholders, and local residents. This multifocal approach enables a more nuanced understanding of the transformations driven by cultural initiatives within local communities and broader territorial systems. By capturing these effects, we can construct a methodological foundation that is not only empirically grounded and replicable, but also adapted to the specific complexities of rural environments, thereby integrating them strategically into the early planning phases of cultural project development.

Designing a cultural project, like any design process approached methodologically, is far too complex to entrust its success to intuition or chance. It demands establishing the solid foundations and guiding principles that guide decision-making and provide a coordinated and clearly predefined framework for action.

These foundations must also incorporate an evaluation logic across all phases of the project. For instance, during the planning stage, tailored survey instruments can be developed to capture the expectations of diverse stakeholder groups. In the implementation stage, key indicators such as participation rates, demographic diversity, and qualitative feedback should be systematically gathered. Following implementation, a combination of methods such as structured interviews, community roundtables, and media impact analyses can be employed to assess actual outcomes in relation to initial objectives.

The use of mixed methods adapted to rural realities is essential. This includes the design of qualitative and quantitative instruments such as:

1. Interviews with cultural promoters and qualified experts
2. Surveys tailored to local populations with simple, closed questions
3. Participatory evaluation formats, such as collective assessments and community feedback sessions

These strategies ensure that project outcomes are interpreted in light of the community's expectations and experiences, reinforcing legitimacy and promoting genuine cultural transformation.

User-Centered Design in the Preliminary Phase of Cultural Projects

Despite the increasing recognition of the importance of culture in rural development, the field of research still lacks a robust methodological approach that contributes to the professionalization of cultural project design. In particular, there is a shortage of methods and tools to support the preliminary phase, an essential moment for defining the scope, feasibility, and community alignment of cultural initiatives. This chapter aims to respond to this gap by outlining the key steps required to proactively involve the local population, promoting agents, and qualified stakeholders in rural areas. These steps include defining objectives, conducting preliminary studies, identifying stakeholders, assessing feasibility and funding opportunities, shaping the conceptual proposal, establishing a time plan, and designing a communication and participation strategy.

To address these challenges, the field of product and service design engineering provides a validated and consolidated theoretical-practical foundation. Particularly user-centered design methodologies such as *human-centered design, design thinking, service design*, or *open innovation* (Nemeth, 2004; Chesbrough, 2006; Brown, 2008; Baines et al., 2017) offer validated frameworks that combine different perspectives within a multidisciplinary approach. These methodologies emphasize participatory and iterative processes, offering tools that are particularly well-suited to the complexity and diversity of rural cultural contexts.

In this context, designing a cultural project can be effectively understood as a form of service design. Both disciplines inherently share a core principle: placing users' needs, expectations, and experiences at the center of the design process, a fundamental premise of human-centered design. Cultural events, much like services, aim to engage participants through meaningful interactions and memorable interactions. This requires a systematic and deliberate approach to planning and organizing people (audiences, staff, collaborators), infrastructures (venues, technology), communication (promotion, interaction strategies, dissemination), and material components (tangible elements such as props, exhibition elements, or stage setups).

Successful cultural project design should draw on methodologies like design thinking and open innovation, leveraging iterative and collaborative approaches to solve complex problems. These methodologies emphasize continuous cycles of prototyping, feedback, and adaptation, ensuring responsiveness to diverse stakeholder needs and the project's evolving context. Incorporating these iterative processes allows cultural initiatives to remain relevant and impactful, but also sustainable and rooted in the lived experience of their communities.

In this evolving landscape, it is also relevant to consider the recent development of Humanity-Centered Design (HCD) as proposed by Donald Norman (2023). While human-centered design focuses on addressing the needs of users within specific contexts, humanity-centered design expands this focus by accounting for broader systemic consequences, long-term societal well-being, and environmental impact. This shift is especially pertinent to rural areas, where cultural projects often intersect with fragile ecosystems, demographic decline, and infrastructural inequalities. Applying a humanity-centered lens encourages designers to reflect not only on immediate user satisfaction, but also on intergenerational equity, community resilience, and sustainability at scale.

Ultimately, adopting a service design framework, complemented by approaches from human-centered design and humanity-centered design, equips cultural project planners with structured, ethical, and context-sensitive methodologies. This approach significantly enhances their potential for success, long-lasting and positive community impact by integrating local voices, reinforcing cultural identity, and promoting social cohesion from the earliest stages of the design process.

Metrics and Indicators in the Preliminary Design Phase for Impact Measurement in Rural Contexts

The development of a cultural and creative sector in rural contexts offers a strategic opportunity to address depopulation, foster personal growth, and promote economic resilience in regions affected by demographic decline. Understanding the cultural sector's impact in these territories supports the formulation of tailored policies and incentives for the rural creative class, reinforcing both cultural life and economic revitalization. Culture and creativity are not merely symbolic forces—they are substantial economic assets that can fuel innovation and regional development (Florida, 2010).

Despite this potential, official data sources such as the Cultural Statistics Yearbook focus primarily on urban metrics and fail to offer indicators specific to rural territories. These statistics typically classify cultural activity by factors such as age, gender, education level, and artistic discipline, but overlook contextual rural nuances, impeding a reliable assessment of cultural impact in these areas.

Effective cultural impact measurement of projects designed for and from rural settings requires the development of specific metrics and context-sensitive indicators during the preliminary design phase. Clearly defined metrics can guide both the formulation and evaluation of projects, aligning them with cultural, economic, environmental, and social sustainability goals. This analytical framework not only enhances the legitimacy of project outcomes but also informs strategic planning and policy innovation in rural creative sectors.

To achieve this, data collection tools must be adaptable to rural settings, where traditional statistics may be scarce or non-existent. Mixed-methods approaches, combining digital or paper-based surveys, focus groups, ethnographic observations, and stakeholder interviews are essential. For instance, Likert-scale surveys can capture shifts in well-being or social cohesion, while in-depth interviews reveal changes in identity, resilience, or economic behavior. New technologies, including mobile apps and participatory GIS, offer dynamic and inclusive solutions, especially in dispersed territories.

A notable precedent is the pilot study: "Claves e Impactos de la Cultura en el Medio Rural" (Sáez Pérez et al., 2024) which emphasizes the relevance of combining perception-based data with evidence gathered through case-study methodology, which is especially suitable when dealing with complex cultural ecosystems. Rather than relying solely on aggregated statistics, the study recommends direct instruments such as semi-structured interviews, questionnaire-based evaluations adapted to rural demographics, and focused discussion groups. These instruments were effectively applied in over 20 case studies across Spain, providing nuanced insights into intangible impacts such as well-being, belonging, and symbolic reputation.

Sustainability-Oriented Impact Dimensions

The indicators proposed below are structured around the three key dimensions of sustainability—social, economic, and environmental—combining qualitative and quantitative evidence. This structure is the result of a thorough review and reinterpretation of the content and contributions from two foundational publications: *Claves e Impactos de la Cultura en el Medio Rural* (Sáez Pérez et al., 2024) and the *Nueva guía para la evaluación de las políticas culturales locales* (FEMP, 2022). Both works offer relevant insights into how to identify and assess the cultural impacts in local contexts.

Building on these references, we present a reorganized and operational synthesis specifically tailored to the preliminary design phase of cultural projects in rural areas. The indicators have been restructured to support a practical and accessible checklist that professionals, researchers, and other stakeholders can use to address how to evaluate sustainability at different levels from the earliest stages of project planning. This approach aims to facilitate the incorporation of impact thinking from the outset, reinforcing the role of the preliminary phase as a critical moment for embedding sustainability criteria into rural cultural initiatives.

Qualitative Impacts

1. **Social Dimension**
   a. *Demographic Impact*
   Cultural projects in rural areas help counter depopulation and increase the attractiveness of these territories, especially for young people and creative newcomers. They reinforce the emotional and identity-based connection of residents to their place of living, stimulating long-term settlement or return migration.
   Indicator: perceived attractiveness of the territory as a place to live and develop a personal or professional project; emotional attachment and intention to remain or return.
   Method for data collection: structured surveys (0–10 scale) to local residents, especially youth and newcomers, along with semi-structured interviews with cultural promoters to collect narratives about population changes, motivations for staying, or returning.
   b. *Community Cohesion and Participation*
   The initiatives strengthen community bonds and foster inclusive participation across age, gender, and origin. Projects promote intergenerational exchange and social cohesion and increase the self-image of the community. The greater the proximity, engagement, and cultural interest, the more positively the impacts are perceived.
   Indicator: degree of social cohesion, perceived quality of interpersonal relationships, and inclusiveness in cultural participation (measured across variables such as age, gender, origin, and social role).
   Method for data collection: community-wide surveys and focus groups that explore relational dynamics, supplemented by participatory mapping and analysis of formal and informal networks activated through the project (e.g., working groups, associations, local collaborations).

2. **Economic Dimension**
   a. *Local Economic Development*
   Cultural projects have a direct impact on local economic activity through increased revenue in hospitality, tourism, retail, and crafts. They also contribute to the reactivation of underused community spaces, helping to sustain social and economic life in small villages and peripheral areas.
   Indicator: perceived change in local economic vitality; reactivation of existing infrastructure (e.g., inns, bars, markets); emergence of new services associated with cultural events.
   Method for data collection: semi-structured interviews with local entrepreneurs and service providers; observational data on commercial activity during events; economic surveys addressed to the community and local businesses.
   b. *Cultural Employment and Entrepreneurship*
   Many projects create professional opportunities for local creators, artists, technicians, and cultural managers. They also encourage the development of new cultural micro-enterprises, valuing local creative skills and promoting sustainable cultural entrepreneurship.
   Indicator: emergence or consolidation of cultural and creative work opportunities; creation of new cultural initiatives or start-ups; level of perceived professional valorization.
   Method for data collection: interviews with local agents and cultural workers; document analysis of project outputs (e.g., contracts, collaborations, new entities); longitudinal monitoring of initiatives launched during or after the project.
3. **Environmental Dimension**
   a. *Impacts on Personal and Community Well-being*
   Cultural initiatives enhance individual and collective well-being. They increase quality of life, boost self-esteem, and foster a deeper connection to the territory, helping people feel better, live more consciously, and enjoy their environment in meaningful ways.
   Indicator: subjective well-being, emotional health, perceived improvement in quality of life and sense of belonging to the local community.
   Method for data collection: standardized surveys with Likert-scale questions on well-being and identity; narrative interviews exploring lifestyle changes, sense of place, and perceived collective transformation.

b. *Reputational Impact*

Projects contribute to a renewed internal and external image of rural areas. Residents report improved perceptions of their own towns, while the broader cultural sector recognizes these territories as innovative, vibrant, and symbolically valuable spaces for cultural production.

Indicator: changes in how the territory is perceived by residents (self-image) and external audiences (media, institutions, visitors); emergence of new narratives or symbolic values.

Method for data collection: mixed-methods approach combining press/media analysis, resident surveys, and interviews with cultural agents to assess changes in visibility, discourse, and symbolic positioning of the locality.

Quantitative Impacts

1. **Social Dimension**
   c. *Demographic Impact*
      - Number of new residents or returnees attracted by cultural projects.
      - Shifts in the age distribution of the local population, especially retention of youth.

        Indicator: demographic change (e.g., increase in young residents, return migration, reduced outmigration).

        Method for data collection: analysis of local registry data (padrones), comparison of pre- and post-project statistics, and surveys exploring migratory intentions and reasons for moving/staying.

   d. *Community Cohesion and Participation*
      - Percentage of the local population involved in project activities.
      - Diversity of participants by age, gender, and background.
      - Number of participatory or co-design initiatives held in the community.

        Indicator: quantified level of community participation; demographic breakdown of engagement; number and type of participatory formats developed.

        Method for data collection: participant records, survey data on frequency and type of involvement, project documentation and community feedback tools (e.g., participatory evaluation sheets).

2. **Economic Dimension**
   c. *Local Economic Development*
      - Measurable increases in revenue in local sectors such as tourism, hospitality, and retail.
      - Number of reactivated services or commercial spaces directly linked to cultural activities.

        Indicator: income variation, increase in service use, number of reopened/activated premises.

        Method for data collection: local administrative and commercial data; direct reports from affected businesses and interviews with project stakeholders.

   d. *Cultural Employment and Entrepreneurship*
      - Number of direct contracts or paid engagements for local cultural professionals.
      - Number of new cultural businesses or projects initiated as a result of these activities.

        Indicator: number of new jobs or contracts created; number of formalized cultural initiatives.

        Method for data collection: internal project records; employment contracts; cultural and local business registries or fiscal records where available; follow-up surveys with entrepreneurs.

3. **Environmental Dimension**
   c. *Impacts on Personal and Community Well-being*
      - Survey results measuring perceived well-being, quality of life, and local attachment.
      - Participation rates in activities promoting sustainable lifestyles or traditional environmental knowledge.

        Indicator: well-being scores and level of involvement in environmentally aware practices.

        Method for data collection: pre- and post-project surveys; attendance tracking of relevant activities; interviews evaluating environmental awareness or lifestyle shifts; comparison between participants and non-participants.

   d. *Reputational Impact*
      - Internal and external evaluations of the territory's image (via scales or qualitative surveys).
      - Media coverage, institutional recognition, and new collaborations generated by the projects.

Indicator: volume and tone of media mentions; qualitative changes in discourse and external partnerships.
Method for data collection: media monitoring (press, digital), content analysis, institutional reports, and surveys of public perception.

Discussion

The integration of user-centered design methodologies in the preliminary phase of cultural project planning offers a tangible opportunity to strengthen the professionalization of the sector in rural contexts. By adapting frameworks such as human-centered design, service design, and design thinking, cultural practitioners can move beyond intuition or tradition and adopt more structured, replicable approaches that prioritize community needs, participation, and impact from the outset. This shift allows for better alignment between cultural actions and the socio-territorial challenges they seek to address, particularly depopulation, loss of identity, and economic fragility.

It is important to consider that the adoption of participatory methodologies not only is a way to improve the quality of cultural projects, but can also generate a deeper change in the way decisions are made. Actively involving the local community from the earliest stages of planning helps to rebalance relations between cultural promoters and citizens, enhancing knowledge, experiences, and desires that are often overlooked. This approach allows communities to overcome vertical or top-down models, instead promoting more shared governance processes, where communities are not only beneficiaries but co-authors of the initiatives. In this sense, participatory design also becomes a tool to strengthen cultural democracy and build stronger and more inclusive social ties.

It is also worth noting that, in the cultural and creative sector, the urgency to implement specific projects or events, combined with the lack of dedicated resources for the preliminary design phase, constitutes a particularly critical challenge. Unlike other professional sectors where the planning stage is formally recognized and remunerated as an integral part of the design and development process, in the cultural field this early phase is often invisible and rarely funded. This absence limits the capacity of cultural agents to conduct thorough diagnostics, design context-sensitive strategies, or engage communities from the outset, ultimately compromising the overall quality and sustainability of the initiative. In this context, the growing recognition of cultural mediation processes as a core component of contemporary cultural practice offers a valuable opportunity to address this gap. Mediators can play a pivotal role in gathering qualitative insights, facilitating early-stage dialogue with local communities, and ensuring that project design is grounded in the lived realities of the territory, thus reinforcing the strategic importance of preliminary data collection and relational work as part of cultural project planning.

At the same time, it is crucial to recognize that participatory processes also involve risks. In some cases, participation can only be requested formally, to give legitimacy to decisions already taken, without the opinions expressed by the communities having any real weight in the final results. In other cases, it can place an excessive burden on local groups that are invited to contribute actively but do not have the resources, time, or expertise to support such involvement. This imbalance is likely to generate mistrust and even a reduction in future participation. To avoid participation becoming an empty gesture or an additional burden for the communities, it is essential to make the community feel valued, through appropriate tools and the recognition of the value of time and commitment of the people involved.

While the proposed methodology is robust and adaptable, its implementation is not without challenges. Rural contexts often face logistical limitations (e.g., digital connectivity, mobility, availability of data), as well as organizational and institutional barriers (e.g., limited funding, insufficient evaluation culture, lack of specialized training). Additionally, participatory methods require time, trust, and local facilitation capacities that are not always readily available. These factors must be considered when transferring or scaling the model to other regions and highlight the importance of embedding the design and evaluation process within local governance and support systems.

The proposed model contributes to bridging the gap between cultural practice and policy, offering a framework that can inform decision-making at multiple levels—from local planning to regional or national cultural strategies. Moreover, it reinforces the role of culture as a cross-cutting driver of sustainable development and territorial cohesion. By articulating specific, measurable, and context-aware indicators, this chapter supports a more evidence-based understanding of cultural impact and lays the groundwork for further academic research and policy experimentation in the field of rural cultural innovation.

Conclusions

What emerges most prominently from this contribution is the pressing need to develop specific methodologies for the preliminary design of cultural events in rural contexts. The current literature highlights the lack of specific tools sensitive to the unique characteristics of each territory. In response, the methodology proposed here adopts a user-centered approach and integrates participatory tools that facilitate meaningful dialogue among citizens, institutions, cultural agents, and the academic sector. The incorporation of impact indicators serves not only to professionalize the cultural sector in low-density areas, but also to enable the creation of more contextually grounded, feasible, and evaluable projects.

Active community participation enhances the effectiveness of cultural initiatives and projects and, more importantly, has the potential to transform decision-making dynamics. When citizens feel heard and their contributions are valued, they become an active part of change. This contributes to the emergence of more equitable and inclusive governance models. However, participation must be carefully structured and resourced. To be meaningful and sustainable, it requires the provision of appropriate tools such as training resources and explicit recognition of the time, effort, and commitment contributed by the community members and the professional mediators involved.

In this context, evaluation instruments play a strategic and multifaceted role. Beyond measuring project outcomes and impact, they also serve as powerful tools for visibility, advocacy, and policy influence. Well-designed evaluation processes can help projects and cultural actors—often overlooked in rural areas—gain recognition, attract funding, and influence institutional agendas. By making the cultural vitality of rural territories visible and measurable, evaluation reinforces the case for culture as a key driver of sustainable development in marginal and low-density regions.

Finally, it is important to reflect on the potential for international transferability of the proposed model. The challenges faced by rural areas in Spain, such as depopulation, lack of services, and diminished institutional presence, are shared by many other European countries, particularly in Southern and Eastern Europe. In this sense, adopting a comparative perspective could open new possibilities: might this model be adaptable to other national contexts? Exploring shared experiences, practices, and tools through transnational cultural networks would contribute to a collective vision of culture as a driver for sustainable development and territorial cohesion at the European level. Strengthening international cooperation through cultural innovation could promote common solutions to shared challenges, reinforcing the strategic role of culture in building more resilient and inclusive rural futures.

References

Baines, T., Ziaee Bigdeli, A., Bustinza, O. F., Shi, V. G., Baldwin, J., & Ridgway, K. (2017). Servitization: Revisiting the state-of-the-art and research priorities. *International Journal of Operations & Production Management, 37*(2), 256–278. https://doi.org/10.1108/IJOPM-06-2015-0312

Brown, T. (2008, June). Design thinking. *Harvard Business Review, 86*(6), 84–92. https://pubmed.ncbi.nlm.nih.gov/18605031/?utm_source=chatgpt.com

Brown, T., & Wyatt, J. (2010). Design thinking for social innovation. *Stanford Social Innovation Review, 8*(1), 30–35. https://doi.org/10.48558/YEN6-4T17. https://ssir.org/articles/entry/the_next_chapter_in_design_for_social_innovation?utm_source=chatgpt.com

Chesbrough, H. W. (2006). The era of open innovation. *Managing Innovation and Change, 127*(3), 34–41.

Cross, N. (2001). Designerly ways of knowing: Design discipline versus design science. *Design Issues, 17*(3), 49–55. https://doi.org/10.1162/074793601750357196. https://direct.mit.edu/desi/article/17/3/49/60062/Designerly-Ways-of-Knowing-Design-Discipline?utm_source=chatgpt.com

European Commission. (2025). *Rural observatory*. https://observatory.rural-vision.europa.eu/?lng=en

Federación Española de Municipios y Provincias. (2022). *Nueva guía para la evaluación de las políticas culturales locales* (2ª ed.). Federación Española de Municipios y Provincias. https://www.femp.es/sites/default/files/multimedia/nuevaguia_indicadoresculturalocal_r.pdf

Florida, R. (2010). *La clase creativa: la transformación de la cultura del trabajo y el ocio en el siglo XXI* (Vol. 117). Grupo Planeta (GBS).

Martinell, A., Heredia Hurtado, M. C., & Vázquez García, L. (Coords.). (2023). *Intuiciones sobre la cultura y la ruralidad: Incorporación al debate sobre la cultura en la Agenda 2030.* Red Española para el Desarrollo Sostenible (REDS / SDSN Spain). https://culturasostenible.org/wp/wp-content/uploads/2023/12/Publicacion_Intuiciones-sobre-la-cultura-y-la-ruralidad-imprimir_v2.pdf

Ministerio de Cultura. (2025). *Temas – MONDIACULT.* https://mondiacult.cultura.gob.es/temas.html

Ministerio para la Transición Ecológica y el Reto Demográfico. (2025a). *Inicio.* https://www.miteco.gob.es/es.html

Ministerio para la Transición Ecológica y el Reto Demográfico. (2025b). *Red Española para el Desarrollo Sostenible (REDS)*. https://www.miteco.gob.es/es/ceneam/recursos/quien-es-quien/red-desarrollo-sostenible.html

Nemeth, C. P. (2004). *Human factors methods for design: Making systems human-centered*. CRC Press. https://doi.org/10.1201/9780203643662

Norman, D. A. (2023). *Design for a better world: Meaningful, sustainable, humanity centered*. MIT Press. https://mitpress.mit.edu/9780262548304/design-for-a-better-world/

Radice, S. (2014). *Designing for participation within cultural heritage: Participatory practices and audience engagement in heritage experience processes*. Tesis doctoral, Politecnico di Milano. https://www.politesi.polimi.it/retrieve/a81cb05a-c437-616b-e053-1605fe0a889a/2014_03_PhD_Radice.pdf?utm_source=chatgpt.com

Sáez Pérez, L. A., Navarro Pérez, M. C., Pinilla Navarro, V. J., & López Azcona, A. (2024). *Claves e impactos de la cultura en el medio rural: El valor de lo intangible*. Ministerio de Cultura. https://culturayciudadania.cultura.gob.es/dam/jcr:784bdae2-e673-4072-9c5a-34cd22277f41/publicaci-n-electrnica%2D%2Dclaves-e-impactos-de-la-cultura-en-el-medio-rural%2D%2Del-valor-de-lo-intengible.pdf

Stickdorn, M., & Schneider, J. (2012). *This is service design thinking: Basics – tools – cases*. John Wiley & Sons. ISBN 978-1-118-15630-8. https://www.wiley.com/en-us/This%2Bis%2BService%2BDesign%2BThinking%3A%2BBasics%2C%2BTools%2C%2BCases-p-9781118156308?utm_source=chatgpt.com

Ulrich, K. T., & Eppinger, S. D. (2016). *Product design and development* (6th ed.). McGraw-Hill Education. https://doi.org/10.1037/e542782016-001

CHAPTER 9

System and Environment: Cultural Impact and Sustainable Conservation in the Case of the Museo della Memoria e dell'Accoglienza

Stefano Magnolo and Ana Galán-Pérez

Introduction

The sustainable conservation of cultural heritage and the construction of collective memory have become key dimensions in contemporary studies on culture and territory. In a context of both global and local transformations, museums not only preserve testimonies and artifacts from the past but also act as active agents in the production of meaning, social cohesion, and sustainable development.

S. Magnolo
Department of Legal Sciences, University of Salento, Salento, Italy
e-mail: stefano.magnolo@unisalento.it

A. Galán-Pérez (✉)
Faculty of Fine Arts, Complutense, University of Madrid, Madrid, Spain

OAAEP Research Group, Funded by the Government of Aragón with FEDER Funds, Zaragoza, Spain
e-mail: am.galan.perez@ucm.es

A. M. Biedermann, F. J. Galán Pérez (eds.), *Interconnections between Culture and Sustainability*,
https://doi.org/10.1007/978-3-032-21061-6_9

In today's global context, this approach aligns with international sustainability policies, particularly those promoted by the United Nations' 2030 Agenda and its Sustainable Development Goals (SDGs), which recognize culture as a fundamental driver of development. International bodies such as UNESCO, the Council of Europe, and the European Union have established regulatory frameworks that reinforce the role of cultural heritage conservation in fostering social cohesion, community resilience, and innovation.

It is within this framework that the present study is situated, offering an analysis of the cultural impact of the *Museo della Memoria e dell'Accoglienza* in Santa Maria al Bagno, located in Nardò (Salento, Italy), as well as its associated indicators and contributions to the sustainable conservation of intangible heritage. This museum serves as a significant space of memory, centered on the reception of thousands of Jewish refugees between 1943 and 1947. Beyond its museographic function, it articulates multiple dimensions of heritage—from the recovery of original murals painted by refugees to the development of a territorial narrative that connects history, landscape, and community. As such, its study is pertinent to reflecting on the role of cultural institutions in heritage sustainability, particularly in relation to their physical, social, and intangible environment.

The main objective of this chapter is to analyze the museum's cultural impact from a systemic perspective (system/environment), considering the interaction between the museum and its surrounding context, and incorporating intangible landscape as a fundamental dimension for evaluating cultural sustainability. The following research questions guide this inquiry: How is the relationship between the museum and its physical and cultural environment manifested? What kinds of cultural and social impacts does the museum generate within the territory and the community?

The methodology adopted is based on a case study developed within the framework of the European Creative Europe program *Culture Moves Europe*, under the title "*Preserving memory through conservation & conversations: towards key concepts for a memorial museum design. Case study: Museo della Memoria e dell'Accoglienza in Santa Maria al Bagno, Nardò (Lecce-Italy)*." The analysis, initiated in 2023, combines interviews with individuals involved in the museum's creation, conservation, and recent management, alongside fieldwork and direct observation, including interviews with groups of visitors.

The case of the *Museo della Memoria e dell'Accoglienza* offers valuable insights into how sustainable conservation and museum practices can foster participatory governance, preserve collective memory through an inclusive approach, and contribute to sustainable development from the

Table 9.1 Research methodology for the period 2023–2025

Phase	*Period*	*Main activities*
Initial approach	May 2023	CroMe, research unit. Initial visit to the museum; meetings with museum leadership and local associations; development of a stakeholder map; contextualization of the museum as a site of collective memory.
Preparation of the culture moves Europe proposal	June 2023–March 2024	Methodological and thematic definition of the project; planning of qualitative research; integration with studies on cultural sustainability (University of Zaragoza).
Project implementation	May–June 2024	Preparation of interviews (in English and Italian) based on research conducted by the cultural sustainability group at the University of Zaragoza: *"Valori immateriali ed esperienza personale nel Museo della Memoria e dell'Accoglienza di Santa Maria al Bagno, Nardò (intangible values and personal experience in the Museum of Memory and Welcome of Santa Maria al Bagno, Nardò) (Lecce-Italy)."* Interviews with key stakeholders (Katia Re, Rivka Cohen, A.P.M.E., Pisacane, Madonna, Taceli); focus group with Israeli visitors; beginning of documentary and bibliographic source analysis. Study of museography, collections, and their conservation.
Project implementation (continued)	May–June 2025	Second phase of fieldwork; in-depth archival analysis; critical interpretation of the museum as a mediator between historical memory and current cultural practices. Study of the safeguarding process, removal, relocation, and conservation-restoration of the murals by Zvi Miller.

perspective of culture. By examining tools used to assess heritage impact, the museum is conceptualized as a living system, capable of engaging with its context and strengthening the cultural sustainability of the territory (Table 9.1).

Theoretical Framework

Cultural Impact and Heritage Sustainability

The preservation, safeguarding, and promotion of cultural heritage enrich the collective human experience and serve as catalysts for peace, inclusion, cohesion, and environmental protection (G20 Ministers of Culture Declaration, Salvador da Bahia, 2024). Ultimately, they represent a pathway toward achieving social and cultural sustainability, governed by international sustainability policies, most notably the United Nations, which launched the 2030 Agenda in 2015, including the 17 Sustainable Development Goals (SDGs). UNESCO, for its part, recognizes and promotes culture as a driver and enabler of sustainable development, advancing the "Culture|2030 Indicators" to measure and highlight culture's contribution to the SDGs.

Since 2019, Europe's Green Deal has echoed the priorities of the 2030 Agenda, emphasizing the cultural dimension of sustainable development within European institutions (European Commission, 2022), and defining and recognizing cultural heritage as a strategic resource for a sustainable Europe:

> Cultural heritage comprises resources inherited from the past in all forms and aspects: tangible, intangible, and digital (both originally digital and digitized), including monuments, sites, landscapes, skills, practices, knowledge, and expressions of human creativity, as well as collections preserved and managed by public or private entities such as museums, libraries, and archives. (2014/C 183/08)

Accordingly, political action and cooperation among EU member states have focused on harnessing the power of culture and cultural diversity to promote social cohesion and well-being, especially through the protection of cultural heritage, as outlined in the European Framework for Action on Cultural Heritage (European Commission, 2018). Within this framework, three main action areas have been identified:

1. Protecting heritage from anthropogenic risks (e.g., illicit trafficking)
2. Promoting sustainable cultural tourism
3. Improving the quality of physical interventions on cultural heritage, especially where poor planning and execution pose a threat

Additional actions aim to safeguard heritage from environmental risks such as natural disasters (2020/C 186/01; CORDIS Results Pack, *Heritage at Risk*, 2028; OMC, *Strengthening Cultural Heritage Resilience for Climate Change*, 2022).

The framework also recognizes the vital role of communities living alongside heritage assets, emphasizing the need for open, participatory, and inclusive processes in heritage conservation and management (OMC,

Participatory Governance of Cultural Heritage, 2018; *Voices of Culture*, 2015).

The social dimension of cultural heritage sustainability is addressed in several Council of Europe documents, most notably the **Framework Convention on the Value of Cultural Heritage for Society**, also known as the **Faro Convention** (CETS No. 199). Article 1 of Section 1 emphasizes that the aim of cultural heritage conservation and sustainable use is human development and quality of life, while also recognizing heritage as a valuable resource for sustainable development.

In terms of shared responsibility, the convention advocates for the development of legal, economic, and professional frameworks that enable joint action, innovation, and democratic participation in the sustainable development of cultural heritage.

Applying this theoretical structure to the study of the Museo della Memoria highlights how a local institution can act as an agent of cultural sustainability. By preserving the memory of Jewish transit following World War II, the museum can contribute to social cohesion, historical reconciliation, and the intergenerational transmission of knowledge, as established in the aforementioned international and European frameworks.

Furthermore, its integration within the coastal environment of southern Italy enables the linkage of memory with environmental sustainability—an essential dimension recognized both in the **Brundtland Report** (United Nations, 1987) and in the European Commission's documents on *heritage at risk* (CORDIS). The museum's activities may also be assessed through the lens of the *Culture|2030 Indicators*, particularly in terms of participatory governance, social inclusion, and sustainable local development.

Additionally, the museum's approach aligns with European initiatives such as **HERIWELL**, **OpenHeritage**, **RURITAGE**, and **Creative Climate Leadership**, reinforcing its potential role in contemporary cultural policies focused on sustainability, resilience, and social innovation. The study of the museum also resonates with Horizon 2020 and Euroreg research projects such as **OpenHeritage**, **CLIC**, **RURITAGE**, **PRO-Heritage**, **ProCultHer**, **Hyperion**, **STORM**, and, particularly regarding participatory governance, **Creative Climate Leadership** (Table 9.2).

European initiatives and best practices in cultural heritage—such as the European Commission's Catalogue of Good Practices and the Council of Europe's Strategy 21—promote three key approaches:

Table 9.2 Main institutions and policy documents on the sustainability of cultural heritage

Institution	*Key documents and initiatives*
United Nations	• *Report of the World Commission on Environment and Development: Our Common Future*- 2015 UN Climate Change Conference (COP21)- 2030 Agenda for Sustainable Development
UNESCO	• Culture 2030
G20 Culture Ministers	• Declaration of G20 Ministers of Culture, Salvador da Bahia, 2024
European Union	• European Green Deal- COM(2022) 709 final- European Commission, 2019–2020/C 186/01–2014/C 183/08- HERIWELL- *Cultural Heritage in Action*- CORDIS *Results Pack: Heritage at Risk*, 2028- OMC, *Strengthening Cultural Heritage Resilience for Climate Change*, 2022- OMC, *Participatory Governance of Cultural Heritage*, 2018- *Voices of Culture*, *Participatory Governance of Cultural Heritage*, 2015- *OpenHeritage*, *CLIC*, *RURITAGE*, *PRO-Heritage*, *ProCultHer*, *Hyperion*, *STORM*- *Creative Climate Leadership*
Council of Europe	• Framework Convention on the Value of Cultural Heritage for Society (Faro Convention), CETS No. 199 • CoE Strategy 21
Government of Spain	• *Green Paper for the Sustainable Management of Cultural Heritage- Final Report of Conclusions*, 2022- *Practical Guide for the Sustainable Management of Heritage*

- **The social dimension**, emphasizing the relationship between heritage, citizenship, and democratic values through participation
- **The economic and territorial dimension**, linking heritage to sustainable local and regional development
- **The educational dimension**, which views heritage as a tool for awareness, education, and research

These lines of action reflect the social mission of conservation, which seeks to contribute to sustainable development in its economic, environmental, social, and cultural dimensions.

In the case of the *Museo della Memoria e dell'Accoglienza*, the focus lies in analyzing and measuring its social and cultural impact, linking its activities to sustainability in these domains and relying on academic research as a fundamental resource for implementing these good practices (Magnolo & Galán-Pérez, 2025a).

Sustainable actions in the social domain integrate equal access to the benefits of cultural heritage, as well as community participation in conservation processes.

Best practices in the conservation and social sustainability of cultural heritage include, on one hand, the **active participation of communities** and their empowerment in defining and managing heritage. This involves recognizing the communities' ownership of heritage and fostering mediation between local communities and public administrations. On the other hand, **genuine participation** implies a transformation in heritage governance, including the redistribution of power and the recognition of rights and ownership. All of this leads toward sustainable governance models that encourage respectful and sustainable production frameworks in which local communities become the main agents and protagonists (Santamarina, 2021).

For the conservation of tangible and intangible heritage to be socially sustainable, a **radical shift in current heritage regimes** is necessary. Protection alone does not guarantee conservation or safeguarding; rather, it is essential to protect and support both the places and the people through active community engagement. This highlights the **collective dimension of heritage**, as emphasized by the SDGs, ensuring cultural and natural biodiversity and promoting the sustainable use of ecosystems and cultural landscapes, while simultaneously reconfiguring traditional power structures, reshaping governance models, and evaluating the extent of participatory governance. This process must include the empowerment and recognition of a plurality of meanings such as territory, memory, knowledge, and identity.

Sustainable cultural actions involve the **preservation of identity**, the **conservation of traditions and cultural practices, innovation in heritage conservation**, and the **promotion of diversity and intercultural dialogue**. In this context, the responsibility to safeguard and transmit cultural heritage to future generations lies not only with public and private entities, but with society as a whole (Magnolo & Galán-Pérez, 2025b; Romero Pastor, 2023). In this way, heritage conservation becomes a dialogue between past and present, seeking solutions that respect the temporality and evolution of cultural forms (Jarauta Marión, 2023).

One of the strongest internal motivations within participatory governance and the advocacy for cultural rights is the need to better protect and preserve cultural heritage. Better protection is pursued, for example, by increasing public awareness of heritage sites and making them more visible, but also through the support of heritage communities, which plays a crucial role in preservation and safeguarding efforts (Galán-Pérez et al., 2023).

To assess the cultural impact of the *Museo della Memoria* from the perspective of sustainable conservation, it is essential to incorporate innovative tools developed at the European level. The **RURITAGE** project proposes a resource ecosystem centered on people and heritage, aligned with the rural context of Nardò. **SoPHIA**, in turn, introduces digital platforms to foster citizen participation and innovation in heritage management (Baioni et al., 2021). The **Heritage Impact Assessment (HIA)** tool allows the anticipation of negative effects on cultural environments by offering a preventive approach (Ashrafi et al., 2022). Additionally, the **"place and memory"** qualitative indicator highlights the symbolic value of heritage spaces marked by migration and exile (Dornelles et al., 2020). The **multidimensional approach proposed by Nocca** (2017) provides robust criteria for integrating social and cultural sustainability into decision-making processes, thereby strengthening the analysis of the museum as a living system within its relational environment.

System and Environment: An Analytical Perspective

The distinction between system and environment has served as a key conceptual tool across multiple theoretical traditions—from cybernetics to General Systems Theory and contemporary sociology (von Bertalanffy, 1968; Luhmann, 1995; Baecker, 2001). Broadly speaking, it enables us to describe how any organized entity—whether an organism, an institution, or a cultural practice—defines its boundaries in relation to a broader and more complex context from which it draws resources and with which it interacts, yet which it can never fully control. Within the framework of social systems theory developed by Niklas Luhmann, this distinction becomes particularly fruitful for social and cultural analysis (Moeller, 2006; Baraldi et al., 1997). According to Luhmann, social systems are autopoietic systems of communication: they constitute and maintain themselves through their own communicative operations, differentiating themselves from their environment, which includes both other systems and the unstructured complexity of the world. In this view, the environment is not merely a physical or geographic "outside," but a dynamic horizon of contingencies and realities that condition the operations of the system.

Applied to the field of culture, this perspective allows us to explore museums as cultural systems that are not merely passive containers of inherited legacies, but active communicative spaces that shape their relationships with communities, territories, and histories through ongoing dialogue with an ever-evolving and expansive environment (Hooper-Greenhill, 2000; Crooke, 2007; Anderson, 2004; Smith, 2006). Social systems theory further enables a distinction between an external environment—comprising natural, psychic, and biological systems—and an internal environment made up of other social systems with which communicative relationships are established (Luhmann, 1995). Within this framework, communities constitute a social environment for museums—understood not only in physical or demographic terms, but as relational spaces of communicative interaction (Watson, 2007).

This approach proves particularly effective for conceptualizing **cultural sustainability**, understood as the capacity of such systems to regenerate over time, respond to new demands, and maintain social relevance (Soini & Dessein, 2016; Dessein et al., 2015). From a systemic perspective, a museum cannot be understood solely as an institution that preserves objects, but as a living system for the production and reproduction of meaning (Hooper-Greenhill, 2000; Anderson, 2004). Its function goes beyond the material presentation of heritage to include ongoing processes of reinterpretation, mediation, and activation of cultural memory (MacDonald, 2013).

The *Museo della Memoria e dell'Accoglienza* in Santa Maria al Bagno offers an emblematic example of this dynamic. It does not merely conserve testimonies from a specific past—the transit of Jewish refugees through southern Italy after World War II—but constructs a communicative space that links this legacy to current sensibilities and debates surrounding hospitality, human rights, and migration (Corsane, 2005; Di Padova, 2018). In this sense, its cultural sustainability depends on its ability to keep open these connections between past and present, between heritage and contemporary communities (Ripp & Rodwell, 2015). The systemic logic allows us to understand this process as a form of **cultural autopoiesis**: the museum generates new communications and new forms of meaning that feed back into its social relevance and strengthen its cultural impact over time (Soini & Birkeland, 2014; Dessein et al., 2015).

From a Luhmannian perspective, a museum's environment is not merely a geographic context but a complex network of social, historical, and territorial relationships (MacDonald, 2013; Smith, 2006). In the case of the *Museo della Memoria e dell'Accoglienza*, this environment consists of the local communities of Santa Maria al Bagno and Nardò, the collective memory of events that shaped the region, and the global flows of memory and meaning tied to experiences of migration and refuge (Assmann, 2011; Ricoeur, 2000). This relational environment is not static, but deeply dynamic. Communities continuously reinterpret their history and relationship with heritage, while the territory itself is re-signified through new social and cultural practices. At the same time, transnational memory networks—including Jewish diasporas and memory tourism circuits—bring new layers of meaning and new interlocutors to the museum (MacDonald, 2013; Logan et al., 2015).

Understanding environment in this way allows us to value cultural sustainability not as the mere conservation of the status quo, but as the capacity of the museum to **creatively engage with these relational processes**, fostering a positive cultural impact that enriches both the system—the museum—and its environment—the communities and territories with which it is connected (Soini & Dessein, 2016).

Intangible Cultural Landscape: The Conservation of Values

The concept of the intangible cultural landscape encompasses the values, memories, practices, beliefs, and symbolic meanings that communities associate with a given territory, going beyond its physical or tangible dimension. Unlike material heritage, this type of landscape emphasizes the intangible aspects that give meaning to a place, as a result of the ongoing interaction between people and their environment over time (Galán-Pérez, Magnolo, 2025).

The Museo della Memoria e dell'Accoglienza highlights the value of the intangible cultural landscape as a central axis for understanding its impact on memory, community, and the environment. This concept—endorsed by UNESCO and the Council of Europe—broadens the heritage approach beyond tangible elements, integrating values, traditions, and connections between people and places as part of a dynamic process. Moreover, the museum's location in Santa Maria al Bagno (Salento coast) is itself a living heritage. Its history as a place of transit and reception (Campo no. 34) is reflected in the environment: the sea, public spaces like squares, and buildings adapted for hosting refugees all contributed to the strengthening of social relationships (Schirosi, 2023).

A key element is the spirit of the place, understood as the symbolic bond between memory and space, which reinforces identity and social cohesion. The museum activates this spirit through its interaction with the territory and the associated memories (Table 9.3 and Table 9.4).

Table 9.3 Elements of the cultural landscape of Santa Maria al Bagno, Nardò (Italy)

Landscape types	*Landscape elements*
Institutional Memory Landscape	Museum in a renovated school building; Garden of Memory
Religious Symbolic Landscape	Palazzo Caputo-Vallone (Synagogue); Casa Rosa, original site of the Zvi Miller murals; brotherhood celebrations and religious processions
Residential Hosting Landscape *(second homes of locals, vacant, donated, or requisitioned)*	Villa Tarantino; Villa Michela Falco (no longer exists); House of Ettore Manno; House of Don Liberato de Metrio; House of Assunta Tarantino; House of Vicenzo Leone Maruccio; House of Giuseppe Cipressa; House of Antonio Vaglio
Economic and Services Landscape	Palazzo Fumarola (for officers and soldiers); Palazzo Nicolini (storage); Paolino Pisacane grocery store
Meeting and Socialization Landscape	Filieri Tavern/Osteria; Gioffreda Bar; Lezzi Tavern/Osteria
Control and Management Landscape	Carabinieri Barracks; Villa De Benedittis (Villa Cristina dei Personé)
Natural Landscape	Natural reserve, Portoselvaggio; Salento coastline

Table 9.4 Images of the cultural landscape of Santa Maria al Bagno, Nardò (Italy)

Museum entrance	Villa De Benedittis (Villa Cristina dei Personè)	Villa Tarantino	Map in the museum

The Concept of Dissonance: The Need for Mediation

Dissonant cultural heritage refers to those heritage elements that generate tensions, conflicts, or contradictory narratives within a community or among different social groups. It is not a homogeneous or consensual heritage, but one that reflects diverse identities, values, and memories that may be in conflict or difficult to reconcile.

This type of heritage reveals aspects of history and culture that are not always comfortable or universally accepted, such as social wounds, displacement, trauma, injustice, or division. As a result, dissonant cultural heritage challenges traditional interpretations and requires mediation processes to foster dialogue, mutual understanding, and the construction of shared memory (Galán Pérez & Vieira, 2023).

In the case of Santa Maria al Bagno, dissonance can be understood in relation to displacement and both institutional and community-based reception. In this context, tensions, conflicts, or divergent narratives emerge from the complex history of migration and hospitality experienced in Nardò. Understanding this dissonance entails analyzing how historical facts and social and cultural relationships have evolved from 1945 to 2025. These tensions are likely to appear in the various ways memory is interpreted, lived, and managed by different stakeholders.

In response to this dissonance, mediation becomes a key tool. First, to promote intercultural dialogue, cooperation, and the building of shared memory between the local community and the Jewish community of refugees and their descendants. Second, to ensure active participation of the communities in the conservation and valorization of heritage, also mediating between them and the institutions. And third, to manage potential disagreements regarding conservation approaches and procedures.

In this framework, mediation not only facilitates interaction and mutual understanding, but also allows for the measurement of the museum's relational impact. Far from being just an archive, the museum emerges as a living social space that fosters encounters, empathy, and symbolic healing. Managing a shared memory requires activating the present between communities and encouraging their continued involvement.

The intangible cultural landscape is the space where these dissonances are expressed and where mediation becomes essential. This concept acknowledges the continuous interaction between people and territory over time, integrating intangible values such as the memory of refugees, local identity, and the symbolic meanings of the place. Beyond the physical environment, it is a symbolically active space where memories, relationships, and historical processes are inscribed.

The protection of these intangible values—such as the memory of hospitality and solidarity—is crucial. Integrating the concept of the intangible cultural landscape enables a holistic vision of heritage, overcoming the divide between material and immaterial aspects and articulating aesthetic, historical, social, and spiritual dimensions.

Analysis: Museum as System and Environment

The Museum as an Active Cultural System

The origin of the museum stems from a civic initiative: the desire to recover the memory of the refugees who passed through Santa Maria al Bagno (1943–1947). In the 1990s, a diverse group of citizens, researchers, and neighbors who had lived through the events promoted the creation of the *Museo della Memoria e dell'Accoglienza*, with the aim of preserving and disseminating the history of the transit camp. This initiative arose despite the initial lack of public administrative support and faced various types of resistance—both administrative and academic (Pisacane et al., 2024). The collection of documents, photographs, letters, personal belongings, and oral histories became the foundation of a museography centered on lived experience (Mennonna, 2006).

The initiative was led by Paolo Pisacane, who—motivated by childhood memories and the discovery of old murals painted by the refugees themselves—promoted the creation of the *Associazione Pro Murales Ebraici* (APME). This association played a key role in gathering tangible and intangible elements, raising public awareness, and applying political pressure that eventually led, decades later, to the museum's opening in 2009. The museum, inaugurated after years of voluntary and persistent management, has become a space for memory education, oriented toward intergenerational dialogue and critical reflection on displacement and reception processes (Schirosi, 2023).

However, the museum's organizational structure has been marked by tensions. Although municipally owned and partially funded by the Puglia Region, the museum was partially appropriated by local authorities, excluding the original promoters of the mural restoration and the final museographic project (Pisacane et al., 2024). A management change occurred in December 2024 through a public tender won by the cooperative *Fluxus*.

Regarding the narrative conveyed in the exhibition, the museum's discourse centers on values such as hospitality, human dignity, and civil resistance against horror. Exhibitory elements combine tangible heritage—like the murals by Zivi Miller, personal belongings of the refugees, and original photographs—with intangible heritage: oral testimonies, family archives, letters, marriage certificates, music, and shared memories that reactivate emotional and intergenerational memory. For Jewish visitors, the museum also represents a reconnection with a part of their family's past during transit through Italy—a period sometimes concealed from descendants due to its proximity to the Holocaust and a desire to overcome trauma (Re, 2024; Cohen, 2024) (Table 9.5).

Table 9.5 Timeline of the restoration of the murals and the opening of the museum

Year/period	*Event*
1943–1947	Operation of the refugee camp in Santa Maria al Bagno
Following decades	Institutional silence and progressive oblivion
1990	Civic initiatives to recover the camp's memory
1995	First awareness campaigns and public petitions for mural conservation
Late 1990s	Creation of the first informative website about the camp; international attention; collection of support signatures; 11-minute documentary aired on RAI
2001–2004	Formation of the *Comitato Recupero dei Murales Ebraici di Santa Maria al Bagno e l'Istituzione del Museo dell'Accoglienza*, later transformed into the *Associazione pro Murales in Santa Maria al Bagno (APME)*, formally recognized in 2004
2003, March 26	Proposal of Law No. 3831 in the *Camera dei Deputati* (2003): Regulations for the restoration of the Jewish "murales" and the creation of the Museum of Hospitality in Santa Maria al Bagno
2004–2005	Awarding of the Gold Medal of Civil Merit to the municipality, promoted by citizen testimonies
2009	Conservation and restoration campaigns for the murals
2009	Opening of the *Museo della Memoria e dell'Accoglienza* and launch of educational activities

Cultural, Social, and Territorial Environment

The *Museo della Memoria e dell'Accoglienza* maintains a dynamic relationship with its cultural, social, and territorial environment, which can be understood as a relational environment in a systemic sense (Luhmann, 1995; Baraldi et al., 1997). This environment includes not only the physical and geographic context of the Salento coastal landscape but also the local communities, collective memories, associative networks, and regulatory and institutional frameworks in which the Museum operates. Within this framework, the intangible cultural landscape—encompassing values, memories, practices, and meanings tied to place—plays a central role (UNESCO, 2013; Ruggles, 2009). The territory of Santa Maria al Bagno is configured as a living heritage site where the memory of the reception of Jewish refugees after World War II is intertwined with contemporary experiences of human mobility and the daily practices of the local community.

Between 1943 and 1947, Santa Maria al Bagno became one of the key reception sites for thousands of Jewish refugees and other displaced persons after the Second World War. A transit camp was established in the area, managed by the British government in collaboration with the United Nations Relief and Rehabilitation Administration (UNRRA). Its legacy has endured not only in documentary records but also in murals, oral histories, and everyday objects that continue to shape a living intangible heritage. For decades, this experience was marginalized in official narratives until various civic initiatives spearheaded its recovery, re-signifying its historical and cultural value (*Search Jews in Italy*, 2021; Ghio & Lelli, 2018).

The relationship between the resident population and the refugees, after an initially complex integration phase, was marked by physical proximity and concrete solidarity. Various oral testimonies collected in Nardò through interviews reveal how local inhabitants provided material and emotional support, integrating the newcomers into everyday village life. Stories of coexistence among children, skill-sharing between women, and the formation of lasting emotional bonds reflect a spontaneous welcoming experience and interfaith coexistence, even in a context marked by resource scarcity and uncertainty. These shared memories have been fundamental to consolidating an inclusive and resilient community narrative (*Comune di Nardò*, 2015, 2016, 2017, 2022).

Today, the Museum acts as a symbolic mediator between the material landscape and the historical narratives it conveys, contributing to the construction of a situated collective memory (Assmann, 2011; Ricoeur, 2000). Activities associated with the museum—educational programs, living memory initiatives, commemorations of concentration camp liberations, and responsible tourism proposals—strengthen this bond between the museum system and its environment, fostering processes of cultural sustainability (Dessein et al., 2015; Soini & Dessein, 2016; Zevi, 2022).

Dialogue Between System and Environment

From a systemic perspective, the dialogue between the museum as an active cultural system and its related environment is expressed through the process of symbolic co-construction (Crooke, 2007; Watson, 2007). Through its integration with local communities, with transnational Jewish memory networks, and with other cultural actors, the museum not only represents the history of the place but also participates in the ongoing re-signification of the territory (Macdonald, 2003, 2013; Logan et al., 2015).

While the cultural and emotional impact on contemporary Jewish communities is evident—being one of the groups most interested in visiting the museum—it is worth analyzing to what extent meaningful exchanges occur with current local communities, how these processes are managed, and what impact they generate. On-site observations and dialogues with associative and academic actors highlight the existence of multiple potential actions that could be implemented but are currently not clearly incorporated into the territory's cultural policies.

In terms of collaborative networks, the museum relies on a framework of relationships with local associations and museums—such as APME, Unitre, and the Jewish Museum of Lecce—national entities like MEIS and the State Archives of Lecce, and international organizations such as Yad Vashem and various Jewish diaspora communities, as well as universities and educational institutions. These networks find expression in pedagogical and outreach activities. Nonetheless, a lack of local involvement persists, which reveals the limits of community ownership of memory (Interviews to Re, Pisacane, Mennona, Tacelli, 2024; *Fondazione per i Beni Culturali Ebraici in Italia*, 2021).

The museum's history is also the history of a process of memory institutionalization. The awarding of the Gold Medal of Civil Merit to the city of Nardò in 2005—partly due to the work of APME—marked a symbolic turning point. In their effort to legitimize the conservation and museographic project, the museum's promoters sent their request to over 170 national and European institutions, receiving favorable responses from figures such as the president of the Italian Republic and the European Parliament's Culture Committee. This institutional backing was crucial in making the museum visible in the public space, although at the local level, cooperation with authorities was not always consistent. The medal's concession was not the result of an official government initiative but rather recognition driven by citizen testimony, illustrating the power of grassroots narratives to influence memory policies. This state recognition enshrines the reception experience as part of the Italian and European moral heritage and legitimizes the refugees' history as a shared legacy.

In this sense, the museum has significant potential to serve as a mediating instrument between different communities—both Jewish and those arising from contemporary migration—facilitating a space for dialogue and the construction of cultural citizenship (Ripp & Rodwell, 2015).

Through its interaction with local and transnational memories, the museum contributes to cultural sustainability but also faces challenges related to community ownership, local stakeholder participation, and the articulation of integrated cultural policies. Its capacity to activate processes of symbolic co-construction and to re-signify the territory represents a promising pathway to strengthen its long-term cultural and social impact (*Comune di Nardò*, 2016; IPSAIC, 2023). (Table 9.6 and Table 9.7).

Table 9.6 Stakeholders involved in the conservation and valorization of the museum

Entity/association	*Contribution to measuring social impact*	*Contribution to measuring cultural impact*
Associazione Pro Murales Ebraici (APME)	Community participation in memory preservation and social education	Promotion of mural conservation efforts and intangible heritage preservation
Archivio di Stato di Lecce	Support for educational activities aimed at raising awareness among new generations	Provision and dissemination of historical documentation
Cooperativa Fluxus	Implementation of programs that strengthen social cohesion and inclusion	Development of exhibitions, cultural activities, guided tours, and educational workshops
Assessorato alla Cultura e Istruzione di Nardò	Integration of the museum into the local educational system, facilitating citizen participation	Promotion of local culture and education about memory
Comune di Nardò (City Council)	Institutional support and promotion of local inclusion	Promotion and facilitation of cultural activities and historical preservation

Table 9.7 Images of the museum and its visitors (2024)

Evaluation of the Museum's Cultural and Social Impact

Sustainability Indicators

Determining what it means for a museum to be "sustainable" in a local context like Nardò (Province of Lecce, Salento region) requires establishing specific margins and indicators tailored to its environment. It is not just about counting visitor numbers or income but about assessing the museum's ability to generate sustained cultural, social, and emotional impact. To do this, three key dimensions are considered: demographic and territorial data, the museum's mission, and levels of community participation, as shown in Table 9.8.

Based on this information, the following applied sustainability thresholds can be proposed (Table 9.9).

The level of sustainability or evaluation scale can be defined based on the degree to which these indicators are met (Table 9.10).

That is, social sustainability is not achieved with a fixed number but rather as a ratio between what is done and the context in which it is done. For Nardò, achieving direct participation of 10% of the population in museum activities—alongside consistent accessibility, inclusion efforts, and cultural production—can be considered a viable and desirable threshold.

To assess the impact of the Jewish communities in Santa Maria al Bagno after WWII—and the descendants who now visit the museum—a transnational, intergenerational, and symbolic view of heritage is required. The goal is not simply quantitative statistical studies, but an understanding of how this place continues to activate identity and emotional ties on a global scale (Table 9.11).

Table 9.8 Data on Santa Maria al Bagno

Contextual Data	*Nardò*
Population	Approximately 30,000 inhabitants
Socioeconomic profile	Coastal area with a strong local identity, migratory tradition, and a partially seasonal tourism-based economy
Cultural infrastructure	Moderate in quantity but high in symbolic value related to memory, human rights, and narratives of reception and displacement

Table 9.9 Quantitative indicators, suggested thresholds, and justification for applied sustainability in the museum

Indicator	*Suggested sustainability threshold*	*Justification*
Local visitors (annually)	≥ 10% of local population (~3000 people)	Significant involvement of the immediate environment
School visitors (local and regional)	≥ 500 students/year	Active presence in the educational environment
Annual community co-created activities	≥ 6 activities/year	At least one every two months with direct participation
Participation of vulnerable groups	≥ 2 specific initiatives/year	Real, not symbolic, inclusion
Local volunteers or collaborators	≥ 15 active people/year	Sign of local ownership and commitment
Local cultural production (exhibits, publications)	≥ 1 original content per semester	Contributes original creativity to the local cultural ecosystem
Presence in European networks/media	≥ 1 active cooperation network (holocaust, memory museums, etc.)	Transnational visibility without losing community roots

Table 9.10 Relationship between indicator compliance and level of social sustainability

Indicator compliance	*Social sustainability level*
≥ 80% indicators met (6–7 of 7)	High (solid and transferable model)
60–79% indicators met (4–5)	Moderate (with potential for improvement)
< 60% indicators met (3 or fewer)	Low (requires active intervention)

Table 9.11 Impact and reach of host communities in the museum

Type of impact	*Reach*
Diasporic cultural impact	The museum serves as a memory hub for communities dispersed across Israel, the U.S., Argentina, and Europe.
Symbolic and identity impact	Connects family memories with a physical location and a narrative of dignity and hospitality.
Testimonial and intergenerational impact	Relatives or descendants recover, reactivate, and transmit history.
Relational impact	The museum keeps diplomatic, cultural, and even touristic ties alive with descendants and associations.

Table 9.12 Quantitative impact indicators

Indicator	*What it measures*	*How it is measured*
Number of international visitors identified as descendants	Direct reach of the diaspora	Visitor logbook, voluntary forms, surveys
Geographical map of families who visited from different parts of the world	Global dispersion and connections	Visual map based on visitor origin data
Number of personal/family testimonies incorporated into the museum	Active participation and intangible legacy	Audiovisual, oral, or textual archive based on interviews
Number of material contributions donated to the museum (documents, objects)	Tangible heritage contribution from families	Archive records and exhibition creation
Number of international Jewish associations linked to the museum	Degree of formal connection with global Jewish community	Agreements, joint projects, shared events
Number of commemorative/ international events	Ritual and symbolic continuity of the bond	Annual agenda, participation in commemorations

Regarding the quantitative impact indicators (Table 9.12 and Table 9.13):

But what about the interaction with local communities?

In studying the relational cultural and social impact, it is essential to establish indicators to evaluate the level of intercultural dialogue, cooperation, and shared memory between communities—that is, how the local community and the Jewish community (both historical and contemporary) interact.

Table 9.13 Qualitative and symbolic impact indicators

Indicator	*What it measures*	*How it is measured*
Sense of belonging expressed in testimonies and networks	Emotional, identity, and spiritual reconnection with the site	Discourse analysis, interviews, written/oral testimonies
Mention of the museum in family and genealogical narratives	Museum's role as a key place for memory transmission	Case studies, published genealogies, biographies, documentaries
Perception of the museum as a "place of healing" or "memory"	Ethical and humanitarian value of remembrance	Perception surveys post-visit, publications, documentaries, social media
Open digital archive of testimonies (collaborative)	Collaborative and participatory dimension of memory	Platform participation, community uploads
Social media and media mentions from the diasporas	Transnational digital visibility of the museum	Hashtag and content analysis, blogs, articles

Below, we present quantitative and qualitative indicators that measure the interaction between both communities in order to assess its intensity, quality, and frequency (Table 9.14).

We consider the joint study of both communities essential for analyzing sustainable social and cultural impact.

This indicator, therefore, helps to interpret the notion of relational heritage not as an object of the past, but as a living process between subjects.

Moreover, it assesses the museum's capacity to foster encounter, empathy, and symbolic reparation. It demonstrates how a community can become the bearer of a memory that is not originally its own, but one it adopts for its ethical and human value.

Study of Identity and Memory: A Personal Questionnaire at the Museum

As part of the research project developed in 2024 on the links between collective memory, identity, and the museum experience, a personal questionnaire was designed and implemented with the aim of exploring the intangible values perceived by visitors to the *Museo della Memoria e dell'Accoglienza* in Santa Maria al Bagno (Nardò, Lecce, Italy).

Table 9.14 Indicators of interaction with local communities

Indicator	*What it measures*	*How it is collected*
Number of intercommunity activities (shared memories, celebrations)	Frequency of events involving both communities	Museum records, local press, annual agendas
Number of joint testimonies	Presence of stories where both communities interact	Oral archive, interviews, publications
Number of symbolic visits or exchanges (Jewish family visits with local hosts)	Ongoing and living hospitality	Guestbook, guide/neighborhood interviews, surveys
Intercultural educational projects	Joint reflections on the past	School programs, collaborations with institutes and memory centers
Perception of the relationship by the local community	Level of pride, identification, or emotional connection to memory	Surveys, focus groups, participatory observation
Perception by Jewish visitors	Recognition of hospitality, respect, and sustained memory	Questionnaires, interviews, social media messages
Perception of respect and trust	Extent to which communities feel respected and valued	Surveys, interviews, focus groups
Shared narratives and intercultural dialogue	Existence of common or enriched discourses between communities	Content analysis of speeches, events, materials
Conflict resolution and handling of differences	Ability to overcome cultural disagreements	Direct observation and testimonies from joint activities
Sense of shared belonging	Identification with the shared space (museum, community)	Open-ended questions in surveys or focus groups
Network mapping to visualize collaborations	Relational intensity and community ties	Visual diagrams based on project and actor interactions

The questionnaire, titled "*Valori immateriali ed esperienza personale nel Museo della Memoria e dell'Accoglienza di Santa Maria al Bagno*," was developed as a qualitative data collection tool for analyzing the participants' subjective experiences. It was administered both in person and online and was structured into five thematic sections:

1. **General information**: age, country of origin, educational background, and motivation for the visit.
2. **Relationship with memory**: perception of the museum's role as a memory space.
3. **Identity dimension**: emotional impact and resonance with one's personal history or cultural values.
4. **Intangible values**: identification of ethical, symbolic, or emotional elements in the museum experience.
5. **Final evaluation and suggestions**: open reflections on the role of the museum in contemporary society.

The **methodological design** took into account the interdisciplinary nature of the approach, integrating contributions from heritage sociology, memory studies, and digital humanities.

In relation to the proposed cultural indicators, the questionnaire focused on the contemporary debate about how to measure the impact of heritage spaces from a more human, qualitative, and integrative perspective. The questionnaire designed and applied at the *Museo della Memoria e dell'Accoglienza* aimed to explore intangible values and the visitor's subjective experience, covering aspects such as identification with the narratives on display, emotional resonance throughout the visit, perception of hospitality, and the link between personal and collective memory.

This proposal reinforces the idea that cultural impact cannot be separated from individual experience or community context. Only through hybrid indicators—capable of integrating the intangible, the emotional, and the symbolic—is it possible to assess the true transformative potential of institutions such as the museum studied.

Currently, the questionnaire is in use (available in Italian and English), and data collection and analysis are expected to take place during a third, ongoing research phase at the museum.

Conclusions: Toward Models of Cultural Sustainability in Memory Landscapes

These reflections reinforce the notion that measuring cultural impact also means capturing the processes of meaning-making, mediation, and appropriation that render heritage a living and transformative dimension.

The case of the *Museo della Memoria e dell'Accoglienza* offers valuable insights into how memory museums can actively contribute to sustainable conservation in diverse contexts. It is not only about preserving a legacy, but about activating dynamic processes of dialogue between past and present, between territory and communities, between local memories and transnational networks.

Through the systemic perspective adopted in this chapter, we have seen how the museum functions as a cultural system in constant interaction with a complex relational environment. This interaction is not merely a contextual backdrop, but the space where meanings, appropriations, and transformations are generated, giving life to the cultural landscape.

The approach developed here highlights several dimensions that can guide future strategies for cultural sustainability:

- The need to anchor memory museums in the intangible cultural landscape of their territory, recognizing the emotional, social, and symbolic bonds that shape it.
- The importance of articulating participatory conservation and management models that strengthen the commitment of both local communities and transnational actors.
- The value of evaluation frameworks that integrate not only quantitative indicators but also qualitative and relational dimensions of cultural impact.
- The potential of memory museums as spaces for intercultural mediation and the construction of citizenship—especially in territories marked by dissonant memories or contemporary migratory dynamics.

From this perspective, the model developed in Santa Maria al Bagno can inspire similar experiences in other contexts, both in Europe and the Mediterranean. Its ability to activate the *spirit of place* and to connect local and global scales makes it a meaningful example of cultural sustainability from the periphery.

There remains an open opportunity to continue exploring these lines in future comparative studies, international cooperation projects, and the development of cultural policies that recognize the value of memory museums as active agents in the cultural and social transformation of territories.

Ultimately, rethinking cultural sustainability through the experience of memory museums means embracing models that not only preserve the past, but activate it as a resource for dialogue, social cohesion, and the construction of more inclusive futures.

References

Anderson, G. (2004). *Reinventing the museum: Historical and contemporary perspectives on the paradigm shift.* AltaMira Press.

Ashrafi, M., Neugebauer, C., & Kloos, M. (2022). Heritage Impact Assessment (HIA): A tool to evaluate the impact of development on cultural heritage.

Assmann, A. (2011). *Cultural memory and Western civilization: Functions, media, archives.* Cambridge University Press.

Baecker, D. (2001). *Wozu Systeme?* Kulturverlag Kadmos.

Baioni, M., Ceschel, A., Demartini, G., Marchegiani, L., Marchiori, M., & Marucci, G. (2021). SoPHIA—Social platform for holistic impact assessment.

Baraldi, C., Corsi, G., & Esposito, E. (1997). *GLU: Glossar zu Niklas Luhmanns Theorie sozialer Systeme.* Suhrkamp.

Camera dei Deputati. (2003, March 26). *Proposta di legge n. 3831: Norme per il recupero dei "murales" ebrei e istituzione del Museo dell'Accoglienza in Santa Maria al Bagno.* https://documenti.camera.it/_dati/leg14/lavori/stampati/pdf/14PDL0044250.pdf

European Comission. (2018). *A New European Agenda for Culture.* https://eur-lex.europa.eu/legal-content/EN/TXT/PDF/?uri=CELEX:52018DC0267

European Commission. (2022). Report from the Commission to the European Parliament, the Council, the European Economic and Social Committee and the Committee of the Regions on the cultural dimension of sustainable development in EU actions (COM(2022) 709 final). https://eur-lex.europa.eu/legal-content/ES/TXT/?uri=CELEX%3A52022DC0709

Comune di Nardò. (2015, December 29). Museo dell'Accoglienza—Santa Maria *al Bagno* [Video]. YouTube. https://www.youtube.com/watch?v=GzDzdjqOiic

Comune di Nardò. (2016, March 15). *CAMPO 34 memorie dell'accoglienza* [Video]. YouTube. https://www.youtube.com/watch?v=GQpPU67_l4o

Comune di Nardò. (2017, March 3). *Campo n.ro 34—Memorie dell'accoglienza—English version* [Video]. YouTube. https://www.youtube.com/watch?v=lvkLsm8VGKY

Comune di Nardò. (2022, January 17). *Il restauro dei Murales Ebraici a Santa Maria Al Bagno (Prima Parte)* [Video]. YouTube. https://www.youtube.com/watch?v=6p_VRPe1gMw

Corsane, G. (Ed.). (2005). *Heritage, museums and galleries: An introductory reader.* Routledge.

Crooke, E. (2007). *Museums and community: Ideas, issues and challenges.* Routledge.

Dessein, J., Soini, K., Fairclough, G., & Horlings, L (Ed.). (2015). *Culture in, for and as sustainable development: Conclusions from the COST action IS1007.* University of Jyväskylä.

Di Padova, F. (2018). *I campi profughi per Jewish displaced persons in Italia tra storia, ricostruzione e memoria, refugee camps for Jewish displaced persons in Italy between history, reconstruction, and memory (1943–1951)* [Doctoral thesis, Università degli Studi di Trieste]. ArTS. https://arts.units.it/handle/11368/29516

Dornelles, A., Gandolfi, E., Mercader-Moyano, P., & Mosquera-Adell, E. (2020). Lugar y memoria como indicador en las intervenciones patrimoniales. Place and memory indicator: Methodology for the formulation of a qualitative indicator, named place and memory, with the intent of contributing to previous works of intervention and restoration of heritage spaces and buildings, in the aspect of sustainability. *Sustainable Cities and Society*, 54.

Fondazione per i Beni Culturali Ebraici in Italia. (2021). *Museo della Memoria e dell'Accoglienza.* https://www.visitjewishitaly.it/listing/museo-della-memoria-e-dellaccoglienza/

G20. (2024). *Declaración de Ministros de Cultura G20, Salvador da Bahia.* https://www.brazilg20.mre.gov.br/

Galán Pérez, A. M., & Vieira, E. (2023). El patrimonio cultural disonante. *Dissonant cultural heritage.* In T. A. i. Güell et al. (Eds.), *Herències / Legacies* (pp. 33–42). Universitat de Barcelona.

Galán-Pérez, A., & Magnolo, S., 2025 (forthcoming). El paisaje cultural inmaterial del Salento: Entre la conservación preventiva, la gestión de riesgos y la sostenibilidad cultural. *The intangible cultural landscape of Salento: Between preventive conservation, risk management, and cultural sustainability.* In *Paisajes Inmateriales.* Editorial Scientifica.

Galán-Pérez, A., Magnolo, S., & Biedermann, A. (2023). Concepto de patrimonio cultural en el contexto de la sostenibilidad. *PH: Boletín del Instituto Andaluz del Patrimonio Histórico, 110*, 322–325. https://doi.org/10.33349/2023.110.5427

Ghio, F., & Lelli, F. (2018). *Guida al Salento ebraico.* Capone Editore.

Hooper-Greenhill, E. (2000). *Museums and the interpretation of visual culture.* Routledge.

IPSAIC—Istituto Pugliese per la Storia dell'Antifascismo e dell'Italia Contemporanea. (2023). *Luoghi della memoria in Puglia.* https://www.luoghidellamemoriapuglia.it/wp-content/uploads/2023/04/13-IPSAIC-PRESENTAZIONE.pdf

Jarauta Marión, F. (2023). El campo expandido de la memoria. En II Simposio de Patrimonio Cultural ICOMOS España (pp. 11–15). Editorial Universitat Politècnica de València (edUPV). https://dialnet.unirioja.es/servlet/articulo?codigo=9469691

Logan, W., Craith, M. N., & Kockel, U. (Eds.). (2015). *A companion to heritage studies*. Wiley-Blackwell.

Luhmann, N. (1995). *Social systems*. Stanford University Press.

Macdonald, S. (2003). Museums, national, postnational and transcultural identities. *Museum and Society, 1*(1), 1–16.

Macdonald, S. (2013). *Memorylands: Heritage and identity in Europe today*. Routledge.

Magnolo, S., & Galán-Pérez, A. (2025a). Diritto, digital humanities e conservazione del patrimonio culturale. In S. Magnolo (Ed.), *Materiali per una sociologia del diritto, della cultura e della conservazione del patrimonio*. Editorial Aracne (Roma).

Magnolo, S., & Galán-Pérez, A. (2025b). Diritto, legittimazione sociale e conservazione critica del patrimonio culturale: Una riflessione sul patrimonio dissonante a partire dal Muro di Berlino. In S. Magnolo (Ed.), *Materiali per una sociologia del diritto, della cultura e della conservazione del patrimonio*. Editorial Aracne (Roma).

Mennonna, M. (2006). *Ebrei a Nardò: Campo Profughi n. 34, Santa Maria al Bagno, 1944–47*. Editore.

Moeller, H.-G. (Ed.). (2006). *Luhmann explained: From souls to systems*. Open Court.

Museo della Memoria e dell'Accoglienza. (2025). *Museo della Memoria e dell'Accoglienza*. https://www.museomemoriaeaccoglienza.com/

Nocca, F. (2017). The role of cultural heritage in sustainable development. *Sustainability, 9*(10), 1882. https://doi.org/10.3390/su9101882

Ricoeur, P. (2000). *La mémoire, l'histoire, l'oubli*. Seuil.

Ripp, M., & Rodwell, D. (2015). The governance of urban heritage. *The Historic Environment, 6*(1), 1–4.

Romero Pastor, J. (2023). El papel de los agentes en la conservación-restauración sostenible del patrimonio cultural. PH: Boletín del Instituto Andaluz del Patrimonio Histórico, 31(110), 311–313. https://www.iaph.es/revistaph/index.php/revistaph/article/view/5417

Ruggles, D. F., & Silverman, H. (Eds.). (2009). *Intangible heritage embodied*. Springer. https://doi.org/10.1007/978-1-4419-0072-2

Santamarina, B. S. (2021). Patrimonio colectivo. Comunidades, participación y sostenibilidad. Buenas prácticas en conservación y revitalización del patrimonio cultural desde los Objetivos del Desarrollo Sostenible (ODS). PH: Boletín del Instituto Andaluz del Patrimonio Histórico, (58). https://www.iaph.es/revistaph/index.php/revistaph/article/view/5000

Schirosi, R. (2023, June 12). *Rinascere in Puglia* [Video]. YouTube. https://www.youtube.com/watch?v=03z1NtT2W9U

Search Jews in Italy. (2021, January 27). *Giornata della Memoria—Storia del Campo d'Accoglienza di Santa Maria al Bagno (Nardò)* [Video]. YouTube. https://www.youtube.com/watch?v=KyNtL_EEeuY

Smith, L. (2006). *Uses of heritage*. Routledge.
Soini, K., & Birkeland, I. (2014). Exploring the scientific discourse on cultural sustainability. *Geoforum, 51*, 213–223.
Soini, K., & Dessein, J. (2016). Culture-sustainability relation. *Sustainability, 8*(2), 167.
UNESCO. (2020). *Indicadores Temáticos Cultura|2030*. https://culture2030.unesco.org/
Voices of Culture. (2015). *Participatory governance of cultural heritage*. https://voicesofculture.eu/2015/06/29/participatory-governance-of-cultural-heritage/
von Bertalanffy, L. (1968). *General system theory*. Braziller.
Watson, S. (Ed.). (2007). *Museums and their communities*. Routledge.
Zevi, L. (2022). *Museo della Memoria e dell'Accoglienza*. https://www.lucazeviarchitetto.it/museo-della-memoria-e-dell-accoglienza.

Interviews

Cohen, R. (June 12, 2024). Personal interview [Stefano Magnolo, Ana Galán-Pérez]. Santa Maria al Bagno, Italy.
Mennona, M. (June 17, 2024). Personal interview [Stefano Magnolo, Ana Galán-Pérez]. Nardò, Italy.
Pisacane, P. (June 17, 2024). Personal interview [Stefano Magnolo, Ana Galán-Pérez]. Nardò, Italy.
Re, K. (June 7, 2024). Personal interview [Stefano Magnolo, Ana Galán-Pérez]. Santa Maria al Bagno, Italy.
Taceli, M.L. (June 15, 2024). Personal interview [Stefano Magnolo, Ana Galán-Pérez]. Nardò, Italy.

CHAPTER 10

Cultural Programming, Social Impact, and Sustainability in Spain's Emptied Areas: The Case of Fuendetodos

Guillermo Juberías Gracia, Eloy Bermejo Malumbres, and Ismael Piazuelo Rodríguez

Guillermo Juberías Gracia's research work has been funded by grants to research groups from public R&D&I organizations in the Principality of Asturias, call for proposals 2024 "Escenarios para el Arte-EsArt" IDE/2024/000731 SEK-25-GRU-GIC-24-05, Asturian Agency for Science, Business Competitiveness, and Innovation (Sekuens).

G. Juberías Gracia (✉)
Department of History of Art and Musicology, Es-Art research team, Universidad de Oviedo, Oviedo, Spain

OAAEP Group, Funded by the Government of Aragón with FEDER Funds, Zaragoza, Spain
e-mail: juberiasguillermo@uniovi.es

E. Bermejo Malumbres • I. Piazuelo Rodríguez
Department of Specific Didactics, Universidad de Zaragoza, ARGOS research team, Zaragoza, Spain
e-mail: e.bermejo@unizar.es; ismaelpiazuelo@unizar.es

A. M. Biedermann, F. J. Galán Pérez (eds.), *Interconnections between Culture and Sustainability*,
https://doi.org/10.1007/978-3-032-21061-6_10

From Depopulated Spain to Revitalized Rural Areas: Challenges and Opportunities for Culture in Rural Spain

In Spain, particularly following the crisis precipitated by the COVID-19 pandemic and the significant housing pressures in major cities and their metropolitan regions, certain rural areas are witnessing an influx of new residents (Del Romero & Fernández Salvador, 2024, pp. 261–286). While this does not yet constitute a substantial migratory movement that could be characterized as an urban exodus, there is a discernible trend toward rural areas as providers of economically, socially, and environmentally sustainable lifestyles. This introduction examines the evolution of the perception of rural culture in Spain over the past decade—from the Empty Spain movement to a rural resurgence—reflected in contemporary cultural representations. We aim to highlight how municipalities with fewer than 1000 inhabitants are increasingly capable of generating sustainable and enriching cultural initiatives that benefit both the territory and its residents.

In recent years, with the emergence of policies aimed at implementing the Sustainable Development Goals (SDGs) outlined in the 2030 Agenda, interest in depopulated rural territories has grown significantly across academic, political, and cultural domains. These SDGs, particularly SDG 4, which is one the focus of this study, should be achieved through the implementation of sustainability policies, defined as "development that meets the needs of the present without compromising the ability of future generations to meet their own needs" (World Commission on Environment and Development, 1987, p. 23). In the Spanish context, scholars such as Gómez Mendoza (2020, pp. 38–70) advocate for promoting a multifunctional model within these territories that encompasses service provision, digitalization, environmental sustainability, a diversified rural economy, and the enhancement of landscape assets-elements notably diverse in Spain.[1]

In this country, the issue of depopulation has received considerable attention over the past decade, particularly under the concept of *España vaciada* "empty Spain," a term popularized by Sergio del Molino (2016) that describes the demographic decline and socio-economic deterioration

[1] In this respect, the Spanish Ministry of Public Works, in the framework of the European Landscape Convention, organized a mapping exercise that resulted in 1263 landscape units throughout the country (Gómez Mendoza, 2020).

affecting extensive areas in the country's interior. Autonomous communities such as Castilla y León, Castilla-La Mancha, and Aragón have been notably impacted by the rural exodus of the twentieth century—a significant migration from rural regions to urban centers—which has left a profound territorial imprint (Collantes & Pinilla, 2011; Pinilla & Sáez, 2021). This phenomenon is not unique to Spain; comparable patterns of rural decline and peripheral marginalization are observed in regions such as France's *Diagonal du vide*—stretching from the Ardennes through the Massif Central to the Pyrenees—as well as in mountainous areas of southern Italy including Basilicata, Molise, and Sardinia, along with certain territories of the former German Democratic Republic (Oliveau & Doignon, 2016; Annunziata et al., 2024; Moorfeld, 2011).

Nevertheless, a growing discourse has emerged concerning the methods of representing, reinterpreting, and revitalizing rural spaces, moving beyond nostalgic or victimizing narratives. This article posits that the case of Fuendetodos can be analyzed within the context of Spain's rural turn over the past decade, which has prompted a cultural and symbolic reassessment of rural environments (Miguélez Carballeira & Usoz de la Fuente, 2024). Within this framework, rural areas have evolved into sites for aesthetic experimentation, collective memory, and social critique. Culturally, this renewed focus on the rural has been explored through various lenses: contemporary literature by authors such as Sara Mesa (*Un amor*, 2020), Juan Gómez Bárcena (*Lo demás es aire*, 2022), and Luis Mario (*Calabobos*, 2025), as well as cinema by Carla Simón (*Alcarràs*, 2022) and Rodrigo Sorogoyen (*As bestas*, 2022). These works provide nuanced perspectives that diverge from idyllic portrayals, highlighting tensions among authenticity, sustainability, and rural gentrification.

Concurrently, some scholars have recently challenged the political and symbolic validity of the term "emptied Spain," citing its stigmatizing and reductive implications. Castelló (2023) advocates replacing this term with the concept of the "resituated rural," which encourages a more dynamic and multifaceted understanding of these territories. From this viewpoint, rural areas are not passive peripheries but active spaces of cultural redefinition, where principles such as sustainability, memory, territorial justice, and gender equity come to the fore.

In Aragón, the autonomous community encompassing Fuendetodos depopulation exhibits a heterogeneous territorial pattern. Although Fuendetodos is part of Zaragoza, the most densely populated province in the region, it lies within Campo de Belchite, one of the most severely affected by demographic decline areas, with a population density of merely 4.4 inhabitants per square kilometer in 2024.[2] The European Union classifies territories with fewer than 8.2 inhabitants per square kilometer as being at severe risk (Gómez Mendoza, 2020, p. 43). Despite having only 145 registered residents in 2024 (INE—National Statistics Institute of Spain, 2024), Fuendetodos attracts approximately 20,000 visitors annually (*El Periódico de Aragón* 2021), highlighting both the appeal of its heritage and the potential for leveraging its cultural assets as engines for development.

The case of Fuendetodos exemplifies a broader trend within the cultural landscape of contemporary rural Spain. Several towns with populations under 1000 residents, situated in regions characterized by low population density, are successfully implementing sustainable cultural initiatives that effectively foster social and economic cohesion in areas traditionally viewed as peripheral or vulnerable to depopulation. A notable example is Urueña, located in the province of Valladolid, which was designated a Book Town in 2007 through the efforts of the Provincial Council of Valladolid.[3] Drawing inspiration from precedents such as Hay-on-Wye in Wales, Urueña has leveraged its modest size—203 inhabitants (INE, 2024)—as a distinctive cultural asset. The municipality hosts twelve bookstores and five museums, including the Music Museum and the Tales Museum, alongside the Joaquín Díaz Foundation, which focuses on the study and promotion of ethnographic heritage. Despite its small scale, Urueña attracted 100,000 visitors in 2024 (García Marbán, 2025).

Another exemplary case is Genalguacil, located in the Genal Valley (Málaga), a region experiencing demographic aging and depopulation. With a population of 397 inhabitants (INE, 2024), the municipality has pursued an ambitious cultural initiative since 1994, earning it recognition as a village-museum.[4] The biennial *Encuentros de Arte* brings together artists from diverse disciplines who reside temporarily to create site-specific works subsequently exhibited in public spaces, resulting in a significant heritage collection. Part of this collection is housed in the Museo de Arte

[2] Data collected in the Summary Document: Estrategia de Desarrollo Local Leader del Campo de Belchite, published in 2024 by the Asociación para el Desarrollo Rural Integral de la Comarca Campo de Belchite (ADECOBEL): https://adecobel.org/wp-content/uploads/2024/01/resumen_EDLL_9DEOCTUBRE.pdf

[3] The official website of this initiative is: https://www.villadellibro.info/

[4] Information on all the artistic programs carried out by Genalguacil can be consulted at: https://pueblomuseo.com/

Contemporáneo Fernando Centeno López. As noted by Casado García and García Díaz (2020), a cornerstone of this model has been the proactive engagement of the local community through a process of collective cultural appropriation.

When evaluating these initiatives, it is essential to consider their contribution to social profitability in addition to economic benefits. Occasionally, certain narratives attempt to economically capitalize on cultural initiatives by assessing their success or failure solely based on profit and loss metrics. Culture often serves as the foundation for a complex transformative process. As Burgos (2020, p. 23) states, "Culture fertilises and vivifies territories, enabling new life plans to come to fruition." A key factor for success in this process, according to cultural manager Lucía Camón, promoter of the platform *Pueblos en Arte* in rural Aragón, is the direct connection between cultural projects and their target audiences. Camón identifies several categories of public in rural areas: permanent residents; *hijos del pueblo* "children of the village," who frequently visit and engage with local activities; holidaymakers who spend extended periods in the village; and visitors who come on a one-time basis (Camón, 2020, pp. 144–145). Empowering local inhabitants in cultural programming is critical for the success of these initiatives. Leveraging the historically strong collaborative spirit within rural communities to sustain year-round cultural activities positions villages as ideal environments for experimentation and artistic creation (Camón, 2020, p. 156).

In the case of Fuendetodos, as detailed below, cultural programming is intricately connected to museum and exhibition institutions that have long-standing traditions within the municipality. Museums situated in small towns possess transformative potential, as recent studies have demonstrated. According to Lorente and Juan (2024, pp. 8-25), institutions as the Museo de Arte Contemporáneo de Genalguacil and the Museo de Dibujo Julio Gavín-Castillo de Larrés (MUDDI) in Huesca operate based on participatory, community-oriented, and contextual frameworks that transcend the conventional museum concept defined by ICOM. These examples demonstrate how contemporary art, rather than being confined to urban centers, can serve as a catalyst for territorial revitalization and symbolic redefinition in rural settings, aligning with the principles of the rural turn.

Fuendetodos: A Small Town Enriched with Cultural Heritage

A distinctive example within the Aragonese and Spanish contexts is the small town of Fuendetodos, located in the province of Zaragoza. This village is renowned as the birthplace of the painter Francisco de Goya y Lucientes (Fuendetodos, 1746—Bordeaux, France, 1828). Unlike other prominent figures in Spanish painting such as Velázquez, Murillo, and Picasso, Goya originated from a modest agricultural village. Consequently, when national recognition of the artist emerged during the latter half of the nineteenth century, efforts were initiated to promote tourism by highlighting his birthplace.

Such initiatives to valorize the birthplaces of eminent artists align with the heritage discourse of nineteenth-century nationalism, which sought to elevate certain individuals to what Peiró (2017) describes as *altares de la patria* "altars of the homeland." A comparable instance can be observed in England with Stratford-upon-Avon, Shakespeare's birthplace. In 1847, a Shakespeare Committee acquired and restored his home to reflect a seventeenth-century English aesthetic. Today, Stratford-upon-Avon serves as a focal point for Shakespearean tourism, significantly shaping and defining its local identity. Comparable examples in Italy include the birthplaces of Michelangelo in Caprese and Leonardo da Vinci in Anchiano, both restored to commemorate the 400th anniversary of Michelangelo's birth in 1875 and the 500th anniversary of Leonardo's birth in 1952, respectively. These sites have undergone processes of memory construction, transforming lived environments into *lieux de mémoire*, as defined by Nora (1984), and have been subsequently musealized.

While this chapter does not aim to provide a comprehensive history of the musealization of Goya's birthplace (*Casa de Goya*)—a topic thoroughly explored in prior studies (Suárez-Zuloaga, 1996; Lozano López, 2008; Castán Chocarro, 2016; Juberías Gracia & Juan García, 2021; Calvo Ruata & Ruyra de Andrade, 2022; Juberías Gracia, 2025)—it is important to highlight how this modest municipality emerged as a symbolic stronghold of Aragonese identity. It has become a pilgrimage destination for contemporary artists and served as the origin of an informal collective of creators that Basque painter Ignacio Zuloaga referred to as the "Fuendetodos pictorial school" (García Guatas, 1976, p. 55).

The identification of Goya's birthplace in Fuendetodos emerged concurrently with the inaugural major exhibition dedicated to the artist in

Spain, organized by the Ministry of Public Instruction and Fine Arts in Madrid in 1900. The preceding year, Goya's remains had been repatriated to Spain from the Chartreuse cemetery in Bordeaux, where they had remained neglected for decades. In 1900, amid the renewed interest in Goya sparked by the Madrid exhibition, journalist Pedro Gascón de Gotor (1900) published an article in the Catalan illustrated magazine *Álbum Salón*. In this article, he transcribed Goya's baptismal certificate housed in the parish church and recounted his 1893 visit to Fuendetodos, including a description of the artist's birthplace. This account was accompanied by an illustration created by his brother, painter Anselmo Gascón de Gotor.

Ignacio Zuloaga, with a keen interest in enhancing the recognition of Goya's legacy, acquired his birthplace in 1915, facilitated by the involvement of both local and international agents. Then he championed the establishment of one of Spain's earliest museums dedicated to photographic reproductions within the *Casa de Goya*, showcasing photographs of the artist's paintings displayed throughout the various rooms of the house (Juberías Gracia, 2025, pp. 244–245). This initiative was ideologically consistent with the objectives of the Institución Libre de Enseñanza, an organization instrumental in advancing pedagogical modernization in Spain during the late nineteenth and early twentieth centuries. Furthermore, Zuloaga personally oversaw the construction of modern schools for the children of Fuendetodos; these facilities currently function as the Ignacio Zuloaga Exhibition Hall (Fig. 10.1).

At the beginning of the twentieth century, the identity of this small town in Zaragoza and therefore its future were linked to the figure of Goya. Fuendetodos became a key point for cultural tourism in Aragon, at a time when this form of tourism was just beginning to take off. In this context, Fuendetodos was visited by the Sindicato de Iniciativa y Propaganda de Aragón (SIPA), an organization established to promote tourism in the region. Since 1925, SIPA organized visits to Fuendetodos and promoted its tourist attractions through its publication, *Aragón* magazine (Juberías Gracia, 2025, pp. 245–247). Despite increasing tourism development in the town, early cultural tourism interest prompted certain enhancements. Consequently, in anticipation of the centenary celebration in 1928, *Aragón* magazine reported on a visit by the provincial architect and outlined planned renovations for the village:

> Construction of a girls' school, Town Hall, and guest house is planned. It is necessary to expropriate certain land parcels and old houses to maximize the

Fig. 10.1 Goya's house and structure of the Museum of Contemporary Engraving, Fuendetodos. (Photographs by Guillermo Juberías)

> visibility of the birthplace of the Immortal Deaf Man [Goya]. On the resulting site plan, a square will be designed with the intention of placing the statue of Goya—originally created by Llaneces and formerly housed in the Prado Museum—at its center, contingent upon securing its transfer to Fuendetodos. Additionally, the Goya Street will be rerouted to provide direct access from the main road, along with its paving improvements.

Although the relocation of José Llaneces's statue to Fuendetodos was not realized, a square was established in front of the *Casa de Goya*. This has led to the creation of a public space that has hosted numerous town festivities and served as a focal point for artistic and cultural activities. A recent example includes the 2018 photography exhibition *Disparate* by Ouka Leele, recipient of the 2005 National Photography Prize. The photographs were taken in this square, featuring residents of Fuendetodos reenacting scenes from Goya's era.

Today, four institutions commemorate the legacy of Goya in Fuendetodos: the *Casa de Goya*, the Ignacio Zuloaga Exhibition Hall, the *Museo del Grabado* (Museum of Engraving), and the Goya Street Art Museum (GSAM). The house where Goya was born suffered looting and destruction during the Spanish Civil War and was subsequently restored in 1946. It was designated as a Historic-Artistic Monument in 1982 and reopened to the public in 1985 (Fig. 10.1). Currently, the house contains limited museographic elements beyond some information about Goya's connections to Fuendetodos; however, recent initiatives, such as Yolanda Picazo's work in 2023, highlight both the potential and opportunity for enhancing its museography.

The Museo del Grabado is situated in close proximity to Goya's birthplace. It occupies a historic residence that was restored in the 1980s with financial support from the Diputación Provincial de Zaragoza and the Fuendetodos Town Council. The museum officially opened its doors in 1989. In the early 2000s, plans were developed for a Museum of Contemporary Engraving, designed by the architectural firm Matos-Castillo, and construction commenced accordingly. The Fuendetodos Town Council owns a collection of 4100 works, representing a significant compilation of graphic art for a town of its size. The museum aimed to establish itself as an international reference center for graphic art and to serve as a hub for research and interpretation in engraving. This institution sought to transform Fuendetodos "into a cultural focal point for new initiatives that would complement its dedicated focus on Goya's legacy and the art of printmaking" (Marcén Guillén, 2010, p. 258). Nevertheless, the financial crisis of 2008 halted further development, leaving the partially constructed structure unfinished; it remains either awaiting completion or demolition and has faced criticism due to its high costs (Sierra, 2017). Paradoxically, this example of unsustainable economic management is set in the midst of a natural landscape in which wind and solar power generators have been installed, renewable energy sources that have transformed the landscape of Fuendetodos (Fig. 10.1).

The Ignacio Zuloaga Exhibition Hall was inaugurated in 1996 within the premises of the former Fuendetodos schools, an institution originally established upon Zuloaga's own initiative. From 1968 to 1981, this site housed the Museo Goya de Fuendetodos, a contemporary art center founded by Federico Torralba, art history professor at Universidad de Zaragoza. This museum featured a collection of donations from artists who created works as tributes to Goya (Lorente Lorente & Picazo Verdejo, 2002, p. 373). However, the quality of these works varied significantly, and the absence of consistent acceptance criteria led to a heterogeneous collection that ultimately hindered the museum's effective operation. Consequently, the museum closed in 1981, coinciding with the closure of Goya's birthplace following approval by the General Directorate of Fine Arts of the Spanish Ministry of Culture for its restoration project. After being released from housing the collection of the defunct Museo Goya, and subsequently stored by the Fuendetodos Town Hall, the former school building has since become a prominent exhibition hall within Zaragoza Province, specializing in graphic art exhibitions (Fig. 10.1). Since its reopening, it has hosted continuous exhibitions through collaborations with prestigious institutions including the Museo Nacional Centro de Arte Contemporáneo Reina Sofía, Calcografía Nacional, Fundación Mapfre, and Musée du Louvre.

In addition, the Goya Street Art Museum (GSAM) was established through the collaboration between the Fundación Goya Fuendetodos and the Zaragoza Chamber of Commerce. This open-air street art museum originated from invitations extended to artists during the annual *Fiesta Goyesca* de Fuendetodos, held every September (Juberías Gracia & Juan García, 2021: 94–97). The inaugural edition occurred in 2017, and since then, numerous artists—including internationally recognized figures such as Xabier Anunzibai (2017) and Boa Mistura (2022)—have contributed murals throughout the streets of Fuendetodos (Fig. 10.2). Several works have significantly influenced the town's visual identity, exemplified by Arantxa Recio's (Harsa) mural located at the San Roque arch near Goya's house-museum.

Cultural Programming in Fuendetodos

The cultural programming in Fuendetodos is examined herein with a focus on its sustainability. The town's cultural management has been largely structured around the Consorcio Cultural Goya-Fuendetodos, an organization established in 1992 with the support of the Diputación Provincial de Zaragoza. This consortium was created to enhance the

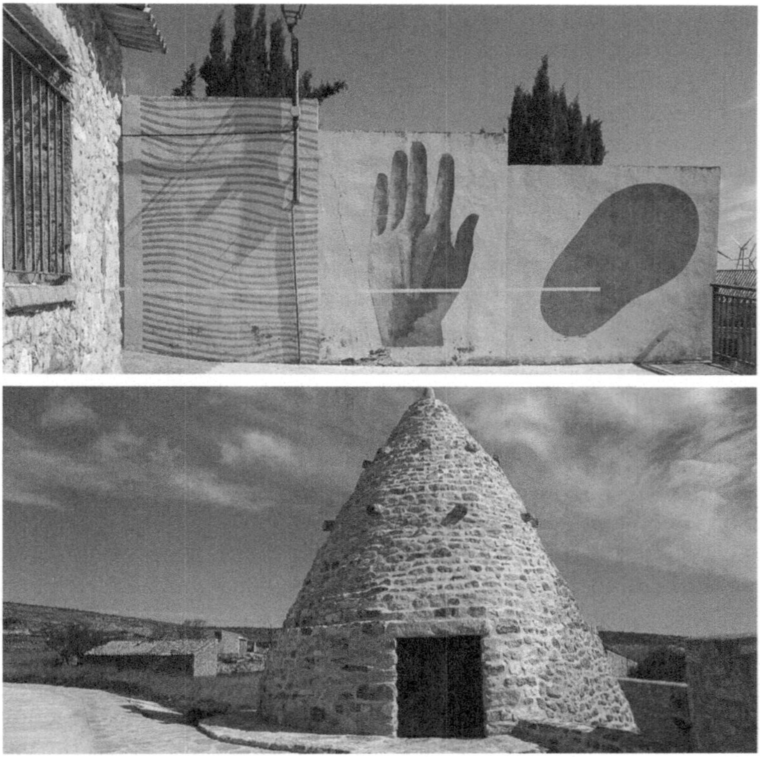

Fig. 10.2 *Agua, mano, vacío*, Xabier Anunzibai, 2017 (above) and Nevero del Calvario (below), Fuendetodos. (Photographs by Guillermo Juberías)

operation of the municipality's cultural venues.[5] As highlighted by Picazo Sánchez (2023), the consortium has facilitated the strengthening of material resources within these spaces and reinforced Goya's role as a central element in local cultural and economic development. Its founding objectives included: (1) establishing Fuendetodos as a hub of cultural activity; (2) delivering a substantial and high-quality range of artistic services and events; (3) ensuring comprehensive management of the *Casa de Goya* and

[5] Its statutes are public and can be consulted on the web page of the Diputación Provincial de Zaragoza: https://www.dpz.es/diputacion/transparencia/informacion-sobre-la-diputacion-provincial/informacion-sobre-normas-e-instituciones-provinciales/ordenanzas-y-reglamentos-propios-1/estatutos-del-consorcio-goya-fuendetodos

the Museo del Grabado; and (4) supporting tourism initiatives alongside emerging cultural industries.

In 1998 the Consortium was expanded to incorporate other municipalities of Zaragoza's province such as Muel, Calatayud, Remolinos, and Cariñena. Subsequently, in 2008, the Fundación Goya Fuendetodos was established as a public entity with the purpose of enhancing cultural promotion associated with the painter's legacy and advancing the development of the proposed Museum of Contemporary Engraving (Blanco Domínguez, 2010, p. 236). By 2010, cultural activity within the municipality had become robust, positioning Fuendetodos as a prominent reference point in redefining Goya's heritage. As museologist Elena Marcén Guillén (2010, p. 260) noted: "the artist's hometown has successfully leveraged its opportunities to create an increasingly ambitious cultural offering that supports Fuendetodos' establishment as a key center for disseminating Goya's legacy," despite the fact that the Museum of Contemporary Engravings—whose collection had already been initiated—was never completed.

Fuendetodos currently offers dynamic and continuous cultural programs throughout the year.[6] This includes guided tours and temporary exhibitions held in the Ignacio Zuloaga Exhibition Hall, complemented by annual events such as the *Romería Goyesca* (Goyaesque pilgrimage) in April and the *Fiesta Goyesca* (Goyaesque festival) in September. Additionally, the Town Council provides three distinct interpretive routes: a monumental route featuring the parish church—recognized as the site of the painter's baptism—along with historic ice deposits (*neveros*) and other ethnological landmarks; a second route focused on the heritage of the ice industry, highlighting *neveros* that historically supplied the city of Zaragoza and are regarded by Domínguez and Ona (1997, p. 73) as indicators of colder, wetter climatic periods (Fig. 10.2); and a third itinerary comprising 23 murals that form the GSAM. Notably, GSAM's free access, integration with the landscape, and ability to create thematic routes align it with sustainable art initiatives found in locations like Genalguacil (Juberías Gracia & Juan García, 2021, p. 98).

The enhancement of Fuendetodos' heritage also pays attention to its natural attractions. The village is situated in an arid landscape that combines steppe and cultivated areas, at a considerable altitude, which ensured

[6] For further information, please consult the Fuendetodos Town Hall website: https://ayuntamientofuendetodos.es/visitas-guiadas/

cold winters that made it possible to obtain ice, an important industry for the municipality. In this respect, the Fuendeverde center exists with the aim of enhancing the natural heritage of Fuendetodos, paying attention to environmental awareness and education.[7] Situated on the site of a former farm within the ruins of an old castle, the facility has been renovated with a strong emphasis on efficiency and sustainability, utilizing local materials, rainwater harvesting systems, biomass heating, and bioclimatic design principles. Its educational approach prioritizes hands-on participation and experiential learning, covering topics such as recycling, renewable energy sources, and responsible consumption. Beyond its permanent exhibition, the center offers environmental trails and workshops focused on botany, geology, and traditional land uses. As part of contemporary local ecological transition strategies, the center aligns with critical perspectives in environmental education. Moreover, in 2022, the Boa Mistura collective contributed to Fuendeverde's development through their intervention within the GSAM project, positioning it as a pivotal point in the local street art circuit. This integration of art, sustainability, and community engagement exemplifies the dynamics of the "rural turn" in contemporary culture (Castelló, 2023), wherein rural settings are increasingly recognized as venues for artistic innovation, alternative educational models, and applied ecological practices.

On the other hand, among the temporary cultural events organized in Fuendetodos, it is worth mentioning the *Fiesta Goyesca*, which represents a participatory event that integrates historical reenactment, the activation of intangible heritage, and community engagement. The portrayal of Goya's birth and baptism, the living pictures, and the revival of traditional Aragonese games form the foundation of a celebration that, similar to other rural contexts, strengthens processes of identity construction (Fig. 10.3). The active participation of both the local community and individuals with emotional ties to the municipality is essential for the project's continuity and legitimacy (Camón, 2020, p. 143). Moreover, while these recreational activities may evoke nostalgia, they also play a significant role in fostering heritage awareness. Community appropriation of heritage largely depends on its social relevance, which can be stimulated through interpretative strategies such as these historical reenactments (Guzmán et al., 2010, pp. 18–20). In addition to these festivities, there is the *Romería Espiritual*, a commemorative walk inspired by an initiative led by

[7] For further information: https://fuendeverde.es/

Fig. 10.3 *Tableaux vivants* representing *Aquelarre*. (Source: Asociación Cultural Goyescos de Fuendetodos)

Ignacio Zuloaga in 1913. That year, accompanied by journalists and intellectuals, the painter journeyed from Zaragoza to Fuendetodos on foot and by cart to protest against rural neglect and to honor Goya's legacy. This march, now referred to as the "spiritual pilgrimage," is annually retraced along a seven-kilometer route. The organizers of the pilgrimage themselves point out how they intend to make a "vindication of the emptied Spain, that which Ignacio Zuloaga valued so much in his works, in which it is also reflected that the essence of the homeland is personified by those who keep its peoples and traditions alive" (*Heraldo de Aragón* & Ayuntamiento de Fuendetodos, 2025).[8]

The Asociación Cultural Goyescos de Fuendetodos plays a pivotal role in both the *Fiesta Goyesca* and the *Romería Espiritual*. Established in 2017 by local residents engaged in historical reenactments, this association has become a significant driver of sustainable cultural revitalization. Its inaugural initiative was a workshop dedicated to creating Goyaesque costumes, led by fourteen women who produced approximately thirty traditional garments representing scenes such as *La Gallinita Ciega*, *El Pelele*, and *El*

[8] The Fundación Zuloaga, a cultural entity whose objectives include the promotion of rural life, has been closely involved in the organization of the Spiritual Pilgrimage: https://fundacionzuloaga.com/lo-local-y-lo-artesanal/

Aquelarre (Fig. 10.3).[9] Since then, the association has launched various projects including the formation of performance troupes and festive characters, ongoing workshops focused on clothing and astronomy, nature trails, and an ambitious holm oak reforestation program initiated during the pandemic. These activities collectively promote cultural sustainability, landscape conservation, intergenerational knowledge transfer, and environmental education. Moreover, these celebrations are part of a long tradition of cultural entertainment, which has its roots in the nineteenth century when Goyaesque *tableaux vivants*, costume parties, and bullfights (*corridas goyescas*) became fashionable (Juberías Gracia, 2022, pp. 365–382).

Additionally, notable is the development of a digital application that integrates both cultural and environmental routes through geolocation, augmented reality (AR), and gamification features. Recent research indicates that this tool enhances visitor engagement by providing clear, interactive information and fostering informal learning; however, it faces limitations due to its exclusive onsite use and relatively low user participation (Bernad Conde, 2020, pp. 81–93).

Educational Uses of Fuendetodos Heritage for Sustainability and Social Development

The cultural activities conducted in Fuendetodos hold significant educational value, not only for local institutions but also for primary and secondary schools across the province of Zaragoza. Accordingly, the following proposals outline sustainable approaches to interpreting Fuendetodos' cultural and natural heritage, with the objective of guiding future initiatives in this domain. Heritage serves as a valuable didactic resource, offering substantial opportunities for emotional, cultural, and civic learning (Fontal, 2003; Castro Calviño & López, 2019). In this context, both the tangible and intangible heritage of Fuendetodos present extensive pedagogical potential applicable across various educational levels—from early childhood education through Baccalaureate—as well as within non-formal education settings.

In the early childhood stage, Fuendetodos, along with its heritage and cultural institutions, constitutes a privileged environment rich in opportunities. In the Spanish curriculum, the *Discovery and Environmental*

[9] For further information: https://goyescosfuendetodos.wordpress.com/

Exploration area capitalizes on pupils' innate curiosity to understand and investigate their surroundings. This domain fosters learning through observation, hands-on interaction, and play, promoting a respectful and conscious relationship with the environment by establishing connections with heritage and nature, thereby contributing to the development of sustainable habits, ecological awareness, and the integration of emotions with knowledge and values from the earliest educational phases (Real Decreto, 95/2022c). Secondly, the *Communication and Representation of Reality* area focuses on language development and various forms of interaction, emphasizing artistic education as a vital mode of expression. This area encourages experimentation with diverse artistic techniques and appreciation of cultural and musical expressions, supporting comprehensive pupil development.

In this context, tailored activities can be developed to leverage the potential of Fuendetodos. For instance, at the Museo del Grabado, a sensory printmaking workshop could be proposed, enabling children to explore stamps, textures, and natural inks. The objective is not to instruct on printmaking techniques, which may be too advanced for children aged 3 to 6, but rather to stimulate their curiosity about mark-making and material experimentation while observing the effects on paper. This activity aligns with the domain of *Communication and Representation of Reality*, fostering artistic expression and enhancing visual-motor coordination. Additionally, the street art featured throughout Fuendetodos offers an opportunity for a "looking game," involving guided routes with stops to examine visual elements such as colors, shapes, animals, or characters depicted in murals. Subsequently, students can collaboratively recreate a mural within the classroom or a secure outdoor area in town using crayons or tempera paints. This process encourages them to express what captured their attention while developing observation skills, visual memory, and creative expression.

At the *Casa de Goya*, an illustrated storytelling activity could be designed to engage pupils with Goya's figure through a tailored narration presented as an illustrated tale, complemented by coloring scenes from the story. Recent research indicates that developing an initial understanding of time through graphic representation during early childhood positively supports the comprehension of historical time in later stages (Sabido Codina et al., 2023).

Furthermore, regarding intangible heritage, pupils can explore local legends or songs via storytelling sessions or dramatized visits. These activities may be enhanced by incorporating gestures or simple instruments. Subsequently, dramatizing the content using puppets promotes communication skills, oral language development, auditory memory, and creativity, thereby creating a connection between cultural expressions within the environment and emotional experiences (Martínez-Rodríguez, 2021).

In primary education, instruction is organized by distinct subjects rather than broad areas. Nonetheless, current legislation encourages interdisciplinary projects that facilitate the simultaneous development of specific competences across multiple subjects (Martínez Rodríguez & Fontal Merillas, 2020). This analysis will concentrate on the subjects most closely associated with its heritage while recognizing its applicability to other disciplines through creative and engaging activities. As noted by Cuenca López et al. (2021), although heritage education is more commonly emphasized in later educational stages, effective practices also exist at the primary level that focus on exploring local heritage, identity, and the environment, thereby fostering responsible citizenship from an early age.

The subjects with the greatest potential for integration in Fuendetodos include *Knowledge of the Natural, Social and Cultural Environment* (sometimes divided into *Social Sciences* and *Natural Sciences*) and *Artistic Education* (occasionally separated into Plastic and Visual Education and Music and Dance). The *Knowledge of the Environment* curriculum aims to develop a critical and sustainable understanding of natural, social, and technological contexts. Students engage in identifying environmental elements, appreciating heritage value, and analyzing human interventions along with their impacts. The study of historical transformations and the role of institutions within society is integral to this discipline. *Arts Education* fosters sensitivity, creativity, and expression through research and practice in visual arts, music, and dance. It also encourages an appreciation of art and cultural heritage as reflections of societal values.

Among the various activities available, a guided visit to the *Casa de Goya* offers valuable insight into his life and historical milieu. Following this, students can develop a chronological timeline and analyze how historical events influenced his work. At the *Museo del Grabado*, after examining Goya's techniques, students have the opportunity to engage in engraving workshops that encourage them to produce artworks addressing contemporary social issues. This approach aligns with innovative strategies for critically reinterpreting artworks, effectively enhancing creativity, empathy, and historical understanding (Bermejo-Malumbres et al., 2024).

The Ignacio Zuloaga Exhibition Hall provides a platform to explore diverse artistic styles and contexts, promoting critical observation and thoughtful expression. Additionally, the urban art scene in Fuendetodos serves as an ideal medium for exploring contemporary art as a form of social commentary or personal expression. Students may collaborate to create a mural at their school conveying their own message. In the Fuendeverde Space, routes can be organized to discover *neveros*, the natural landscape or fortifications, reflecting on their historical function, their integration into their surroundings and their human impact, thus promoting environmental and heritage awareness. Intangible heritage can be addressed through interviews, dramatizations or recordings of older people, exploring traditional sayings, customs, or celebrations. Pupils could illustrate a "book of folk wisdom" or act out one of these traditions in a group.

For Secondary Education (ESO), in accordance with Real Dereto 217/2022a, it is appropriate to maintain the division by cycles. The subject of *Geography and History* aims to foster students' critical understanding of historical time, geographical space, and contemporary society, promoting active and engaged citizenship. This educational approach is based on a competency and context-oriented framework that integrates source analysis, collective memory, and cultural heritage. Numerous studies support the educational value of heritage as a resource for developing historical thinking and critical awareness. Felices de la Fuente et al. (2020) emphasize that local heritage facilitates meaningful learning experiences, while Gómez Carrasco et al. (2020) highlight the effectiveness of using heritage centers in History teaching to activate active methodologies. Pinto and Ibañez-Etxeberria (2018) advocate for its potential to foster historical empathy and build inclusive identities. Serrano Andrés and Sebastián López (2024) propose that the *Geography and History* subject can serve as an instrument to combat depopulation and encourage youth attachment through projects linked to the local environment; they also underscore the necessity of reinterpreting rural areas as spaces with future potential, aligning with the rural turn discussed in this article.

During the first cycle (the first two years of secondary education), general methodologies for acquiring historical and geographical knowledge are introduced. The most suitable content areas for exploring the heritage of Fuendetodos include, firstly, *Block A: Challenges of the Contemporary World*, which addresses fundamental concepts related to geographical, social, and cultural diversity, source interpretation, heritage education, as well as geographical space and its representation. Secondly, *Block B: Societies and Territories* fosters interest in understanding pre-industrial societies, art as a historical source, and the collective memory and legacy of communities.

Fuendetodos provides diverse educational opportunities: the *Casa de Goya* offers insights into daily life in the eighteenth century and presents the artist as a significant historical figure; the Museum of Engraving facilitates critical image analysis as a means of historical interpretation, potentially culminating in period-based narratives. Additionally, the *neveros* and castle serve as sites for reflection on resource exploitation and territorial organization. Within the Fuendeverde Space, students can engage with field notebooks and conduct analyses concerning landscape changes and society-nature relationships in alignment with SDGs. These activities promote competences such as source interpretation, contextual understanding of present circumstances through historical perspectives, and critical evaluation of heritage.

For the last two years of ESO, the curriculum advances in complexity, enabling a deeper level of critical analysis. *Block A* focuses on cultural heritage and its preservation, collective identity, and active citizenship. *Block B* covers topics such as the Enlightenment, the French Revolution, the War of Independence, and art from the eighteenth to the twentieth centuries. This cycle maximizes resources like the *Casa de Goya* by connecting it to the context of the Enlightenment, early liberal movements, and the War of Independence. At the *Museo del Grabado*, students have the opportunity to analyze a work by Goya and contemplate its social or political message. Subsequently, they draw parallels between this content and contemporary issues by composing editorial-style texts or creating original graphic works that convey a critical perspective. This activity is designed to foster historical thinking, artistic education, and social awareness. Additionally, spaces such as the Ignacio Zuloaga Exhibition Hall or urban art installations encourage reflection on social representations, art as a form of protest, and memory's role in shaping contemporary history.

At the Baccalaureate level, the structure and curricular framework established by Real Decreto 243/2022b emphasize the importance of connecting academic content with contemporary social challenges, fostering critical analysis of historical events, and promoting an active, reflective, and culturally engaged citizenship. Subjects within the Humanities and Social Sciences pathway—such as *History of the Contemporary World*, *History of Spain*, *Art History*, and *Geography*—are particularly well-suited to developing competences related to the heritage of Fuendetodos. These competences include critically analyzing the past to better understand the present, interpreting historical and artistic sources, recognizing art as both an ideological and social document, applying geographic methodologies to examine territorial, demographic, and environmental dynamics, and appreciating heritage as a cultural construct that shapes collective identity (Álvarez-Domínguez et al., 2017).

In the *History of the Contemporary World* course, the study of Goya's life and work provides an opportunity to explore key developments of the eighteenth and nineteenth centuries, including the crisis of the *Ancien Régime*, the French Revolution, the War of Independence, and Enlightenment thought. Through visits to Goya's House and the Museum of Engraving, students can develop a mind map that links his engravings and personal experiences with these significant historical events. A concluding debate is proposed to consider whether Goya would be regarded as an activist in today's context. Within *Spanish History*, Goya's legacy serves as a reference point for understanding the formation of the liberal state, socio-political changes during the nineteenth and early twentieth centuries, as well as issues related to historical and democratic memory. In *Art History*, resources such as the *Museo del Grabado*, Ignacio Zuloaga Exhibition Hall, and street art provide valuable material for analyzing diverse artistic languages as forms of symbolic, ideological, and political expression. Students are encouraged to examine stylistic evolution, techniques, and thematic content while appreciating art's role as both social critique and historical documentation. In *Geography*, locations like Fuendeverde Space, local *neveros*, and rural environments facilitate interpretation of territory as a product of historical and social processes. These settings also prompt reflection on sustainability and landscape transformation. An illustrative activity could involve creating a thematic map that correlates physical geography with human factors in relation to contemporary challenges such as depopulation, climate change, or land use management.

Finally, the possibilities of non-formal and informal education are even more than in the regulated stages, as they are not subordinated to an official curriculum. Given the diversity of possible audiences, different activities adapted to different groups are proposed. For older people, a workshop is proposed at the *Museo del Grabado* which, after a visit centered on Goya's most expressive works, invites them to share memories linked to emotions and to create an engraving that represents them, promoting memory, artistic expression, and biographical recognition. For families with children, a gymkhana through the town is proposed with tests adapted to all ages, ending with a collective workshop in the Espacio Fuendeverde. This gamified activity encourages intergenerational learning and makes the visit more dynamic. Finally, for groups interested in Goya, we suggest a bilingual guided tour focusing on his local context, complemented with an introduction to engraving and a debate on the validity of his critical thinking. This proposal adapts to different cultural levels and allows working on concepts such as change and historical continuity (Piazuelo Rodríguez & Rubio Navarro, 2024; Paricio, 2018).

Conclusions

Despite the fact that the population density in the Campo de Belchite remains critical and that the current population of Fuendetodos represents only 22% of what it was in 1930 (663 inhabitants, INE, 1930), recent years have seen a certain demographic stabilization that allows for more sustainable prospects. Although the municipality no longer has its own school, the social fabric remains active thanks to an intense and diverse cultural program, in which community involvement is fundamental, as highlighted throughout this article. Organizations such as the Asociación Cultural Goyescos de Fuendetodos exemplify a model of community participation that includes both residents and descendants of the town, strengthening identity ties and collective commitment to the preservation and reinterpretation of local heritage.

From this perspective, Fuendetodos serves as a paradigmatic example of the rural turn analyzed in the introduction: a symbolic, cultural, and political shift that reinterprets rural areas as active spaces for cultural and social production. The local community has been able to generate its own discourse on rurality, activating processes of heritage reappropriation that enrich the town's cultural life and contribute to its resilience.

Regarding its cultural facilities—particularly the Casa de Goya, the Museum of Engraving, and the Ignacio Zuloaga Exhibition Hall—although the latter carries out an outstanding exhibition activity—greater institutional involvement (at municipal, provincial, and regional levels) is essential to ensure a more dynamic and sustainable management. The development of contemporary communication strategies capable of attracting new audiences and enhancing the external visibility of

Fuendetodos' cultural project is key to consolidating its economic, social, and symbolic sustainability.

Furthermore, Fuendetodos' heritage and educational potential extends beyond its evident tourism, curricular, or cultural possibilities. Heritage can become a strategic resource against the effects of depopulation, which not only entails population loss but also environmental degradation and disruption of local cultures. From this perspective, critical heritage education—in formal, non-formal, and informal contexts—emerges as a crucial tool to promote community cohesion, intergenerational transmission, and strengthen rural social fabric. This aligns with SDG 4 (Quality Education) and SDG 11 (Sustainable Cities and Communities) (Serrano Andrés & Sebastián López, 2024). Educational proposals that incorporate the analysis of the landscape, traditional architecture, historical water management systems, and the evolution of rural ecosystems also align with SDG 13 (Climate Action) and SDG 15 (Life on Land). Additionally, addressing artistic expressions such as Goya's engravings or contemporary urban art facilitates engagement with content related to SDG 16 (Peace, Justice, and Strong Institutions) by promoting critical thinking, intercultural dialogue, and awareness of the symbolic dimensions of social justice (Piazuelo Rodríguez & Rubio Navarro, 2024). Ultimately, Fuendetodos should be understood not only as a tourist or educational site but as a living rural territory that is redefined and oriented toward the future.

References

Álvarez-Domínguez, P., Martínez-Valcárcel, N., & García-Marín, R. (2017). El patrimonio cultural en los recuerdos del alumnado al finalizar el Bachillerato en España: educación e identidad patrimonial. *Revista Tempo e Argumento, 9*(22), 198–235. https://doi.org/10.5965/2175180309222017198

Annunziata, A., Scorza, F., et al. (2024). Unveiling intra-rural divides: Investigating decline and prosperity in rural areas. The Case Study of Southern Italy. *European Planning Studies, 32*(7), 1478–1505. https://doi.org/10.1080/09654313.2024.2335312

Bernad Conde, M. S. (2020). Nuevas tecnologías y difusión del turismo cultural: descubriendo a Goya con realidad aumentada. *ROTUR, Revista de Ocio y Turismo*, 14(1), 81–93. https://doi.org/10.17979/rotur.2020.14.1.5945

Bermejo-Malumbres, E., Piazuelo-Rodríguez, I., Hernández de la Cruz, J. M., & González-González, J. M. (2024). La recreación de obras de arte en el aula para la formación de futuros docentes. Interpretación crítica y transformación contemporánea. *Her&Mus. Heritage & Museography, 25*, 167–183. https://doi.org/10.60940/hermusv25id432220

Blanco Domínguez, P. (2010). Aproximación a la memoria reciente de Goya en Aragón (1979–2011). *Artigrama, 25*, 221–238. https://doi.org/10.26754/ojs_artigrama/artigrama.2010257560

Burgos, B. (Ed.). (2020). *Pensar y hacer en el medio rural: Prácticas culturales en contexto.* Ministerio de Cultura.

Calvo Ruata, J. I., & Ruyra de Andrade, M. (Eds.). (2022). *Zuloaga, Goya y Aragón: la fuerza del carácter.* Ayuntamiento de Zaragoza.

Camón, L. (2020). De los públicos a las comunidades. In B. Burgos (Ed.), *Pensar y hacer en el medio rural. Prácticas culturales en contexto* (pp. 143–157). Ministerio de Cultura.

Casado García, C., & García Díaz, M. (2020). Genalguacil, el museo habitado. *Diferents. Revista de museus, 5*, 110–127. https://doi.org/10.6035/Diferents.2020.5.7

Castán Chocarro, A. (2016). *Señas de identidad. Pintura y regionalismo en Aragón (1898-1939).* Institución Fernando el Católico.

Castelló, E. (2023). The resituated rural: Exploring narratives beyond the empty Spain. *Journal of Spanish Cultural Studies, 24*(4), 529–544.

Castro Calviño, L., & López, F. R. (2019). Educación patrimonial a partir do territorio próximo. In X. C. Macía Arce, F. X. Armas Quintá, & F. Rodríguez Lestegás (coords), *La reconfiguración del medio rural en la sociedad de la información: Nuevos desafíos en la educación geográfica* (pp. 699–711). Andavira Editora.

Collantes, F., & Pinilla, V. (2011). *Peaceful surrender: The depopulation of rural Spain in the twentieth century.* Cambridge Scholars Publishing.

Cuenca López, J. M., Martín Cáceres, M. J., & Estepa Giménez, J. (2021). Teacher training in heritage education: Good practices for citizenship education. *Humanities and Social Sciences Communications, 8*(62), 1–8. https://doi.org/10.1057/s41599-021-00745-6

Domínguez, J. A. & Ona, J. L. (1997). Fuendetodos: los paisajes que Goya conoció. *Flora Montiberica*, 6, 72–75.

Del Romero, R. L., & Fernández Salvador, N. (2024). Looking for Thoreau's cabin. Post-pandemic lifestyle migrations and rural idyll in alt Millars (Castellón, Spain). *AGER: Revista de Estudios sobre Despoblación y Desarrollo Rural (Journal of Depopulation and Rural Development Studies), 39*, 261–286. https://doi.org/10.4422/ager.2024.08

Del Molino, S. (2016). *La España Vacía: viaje por un país que nunca fue.* Turner.

El Periódico de Aragón. (2021). Fuendetodos, viaje a la patria de Goya. El Periódico de Aragón. Retrieved June 19, 2025, from https://www.elperiodicodearagon.com/el-mostrador/2021/03/29/fuendetodos-viaje-patria-goya-46449128.html

Felices de la Fuente M del M., Chaparro Sainz, Á., & Rodríguez Pérez, R. A. (2020). Perceptions on the use of heritage to teach history in secondary education teachers in training. *Humanities and Social Sciences Communications, 7*(123), 1–10. https://doi.org/10.1057/s41599-020-00619-3

Fontal, O. (2003). *La educación patrimonial: teoría y práctica para el aula, el museo e Internet.* Trea.

García Guatas, M. S. (1976). *Pintura y arte aragonés (1885–1951).* Librería General.

García Marbán, M. (2025). Urueña recibe cerca de cien mil visitantes en 2024. El Norte de Castilla. Retrieved June 19, 2025, from https://www.elnortedecastilla.es/valladolid/provincia/uruena-recibe-cerca-cien-mil-visitantes-2024-20250109105431-nt.html

Gascón de Gotor, P. (1900). Goya. Álbum Salón (01-01-1900): 114–115.

Gómez Carrasco, C. J., Miralles Martínez, P., Fontal, O., & Ibañez Etxeberria, A. (2020). Cultural heritage and methodological approaches. An analysis through initial training of history teachers (Spain–England). *Sustainability, 12*(3), 1–21. https://doi.org/10.3390/su12030933

Gómez Mendoza, J. (2020). Despoblación, territorio y sostenibilidad (ODS). *Revista de estudios Locales—CUNAL* 232: 38–70.

Guzmán, A., Fernández, G., & Valenzuela, S. (2010). La recreación histórica como herramienta de la interpretación para recuperar y preservar el patrimonio en peligro. *Boletín de interpretación, 23*, 18–20. https://doi.org/10.1080/14636204.2023.2272044

Heraldo de Aragón & Ayuntamiento de Fuendetodos (2025). IV Romería Ignacio Zuloaga, hasta el corazón de los orígenes de Goya. Heraldo de Aragón. Retrieved June 20, 2025, from https://www.heraldo.es/branded/iv-romeria-ignacio-zuloaga-hasta-el-corazon-de-los-origenes-de-goya/.

INE - National Statistics Institute of Spain. (2024). *Cifras oficiales de población de los municipios españoles en aplicación de la Ley de Bases del Régimen Local.* Fuendetodos (Zaragoza). Madrid.

Juberías Gracia, G. (2022). Bailarinas como eternas majas: el imaginario goyesco y los trasvases entre danza, pintura y cuadros vivos. In I. Murga Castro et al. (Eds.), *Tras los pasos de la Sílfide: Imaginarios españoles del ballet romántico a la danza moderna* (pp. 365–382). Ministerio de Cultura.

Juberías Gracia, G. (2025). *Culto a los maestros e identidad nacional: Goya como referente para la cultura visual española (1868–1928).* Institución Fernando el Católico.

Juberías Gracia, G., & Juan García, N. (2021). Identity, memory and rural development: Public art in Fuendetodos (Zaragoza, Spain). *Terra, revista de desarrollo local, 9*, 81–102. https://doi.org/10.7203/terra.9.22076

Lorente Lorente, J. P., & Picazo Verdejo, E. (2002). El «Museo Goya» en Fuendetodos (1968–1981). *Su colección de arte contemporáneo. Seminario de Arte Aragonés, XLIX-L*, 367–404.

Lorente Lorente, J. P. & Juan García, N. (2024). El nombre no hace la cosa. Nomenclaturas/taxonomías de los museos de arte contemporáneo en la España rural. *Diferents, Revista de museos, 9*, 8–25.

Lozano López, J. C. (Ed.). (2008). *La memoria de Goya (1828–1878)*. Gobierno de Aragón.

Marcén Guillén, E. (2010). El Espacio Goya de Zaragoza: historia de un proyecto y propuestas arquitectónicas (1998–2011). *Artigrama, 25*, 239–261.

Martínez Rodríguez, M., & Fontal Merillas, O. (2020). Dealing with heritage as curricular content in Spain's primary education. *The Curriculum Journal, 31*, 77–96. https://doi.org/10.1002/curj.7

Miguélez Carballeira, H., & Usoz de la Fuente, M. (2024). Cultural engagements with the rural in contemporary Spain: Manifestations and critical tensions. *Journal of Spanish Cultural Studies, 25*(4), 465–475. https://doi.org/10.1080/14636204.2024.2424614

Martínez-Rodríguez M. (2021). El patrimonio a través de la Educación Musical: Tratamiento y enfoque en el currículo de Educación Primaria. *Revista Electrónica Complutense de Investigación en Educación Musical - RECIEM, 18*, 27-37. https://doi.org/10.5209/reciem.686

Moorfeld, M. (2011). Landscapes in eastern Germany at a turning point: Linkages between population decline, ageing and land consumption. *European Countryside, 3*(2), 29–50. https://doi.org/10.2478/v10091-011-0007

Nora, P. (1984). *Les lieux de mémoire*. Éditions Gallimard.

Oliveau, S., & Doignon, Y. (2016). La diagonale se vide ? Analyse spatiale exploratoire des décroissances démographiques en France métropolitaine depuis 50 ans. Cybergeo: European. *Journal of Geography, 763*. https://doi.org/10.4000/cybergeo.27439

Paricio, J. (2018). El valor de la historia. Estudio de alternativas curriculares en Secundaria (1): Una visión integrada de las transformaciones (cambio/continuidad) que conducen y modelan el presente (conciencia histórica). *Clío: History and History Teaching, 44*, 232–247.

Peiró Martín, I. (2017). *En los altares de la patria: La construcción de la cultura nacional española*. Akal.

Piazuelo Rodríguez, I., & Rubio Navarro, A. (2024). Cambio y continuidad en el concepto de belleza: Una propuesta para educar en pensamiento histórico con el museo. *Clío. History and History Teaching, 50*, 381–402. https://doi.org/10.26754/ojs_clio/clio.20245042019

Picazo Sánchez, Y. (2023). *Plan Estratégico "Goya-Fuendetodos: una nueva mirada", Trabajo Final de Máster*. Universitat de Barcelona. Retrieved June 20, 2025, from https://diposit.ub.edu/dspace/handle/2445/207767

Pinilla, V., & Sáez, L. A. (Eds.). (2021). *La despoblación de la España interior*. Prensas de la Universidad de Zaragoza.

Pinto, H., & Ibañez-Etxeberria, A. (2018). Constructing historical thinking and inclusive identities: Analysis of heritage education activities. *History Education Research Journal, 15*(2), 342–354. https://doi.org/10.18546/HERJ.15.2.13

Real Decreto. (217/2022a). de 29 de marzo, por el que se establece la ordenación y las enseñanzas mínimas de la Educación Secundaria Obligatoria.

Real Decreto. (243/2022b). de 5 de abril, por el que se establecen la ordenación y las enseñanzas mínimas del Bachillerato.

Real Decreto. (95/2022c). de 1 de febrero, por el que se establece la ordenación y las enseñanzas mínimas de la Educación Infantil.

Sabido Codina, J., Bellatti, I., & Callarisa, J. (2023). Didáctica del tiempo en educación infantil. El tiempo percibido de los infantes 3-6 años a partir del dibujo libre. *Revista Electrónica Interuniversitaria de Formación del Profesorado, 26*(1), 141–156. https://doi.org/10.6018/reifop.532171

Serrano Andrés, A., & Sebastián López, M. (2024). Estrategias educativas para la lucha contra la despoblación: Una propuesta transversal a través de las asignaturas de Geografía e Historia en institutos rurales de Aragón. In F. J. García-Delgado et al. (Eds.), *Con(textos) de desarrollo local. Dos décadas de Geografía aplicada (2003–2025)*. (pp. 317–338). Asociación Española de Geografía (AGE). https://doi.org/10.21138/dl.2024.17

Sierra, A. (2017). El museo de Goya, donde las grietas dibujan un cuadro de despilfarro. El Español. Retrieved June 20, 2025, from https://www.elespanol.com/espana/20170121/187731750_0.html

Suárez-Zuloaga, M. R. (Ed.). (1996). *Zuloaga en Fuendetodos*. Diputación Provincial de Zaragoza.

World Commission on Environment and Development (1987) *Our common future*. United Nations.

CHAPTER 11

Art Suspended over the Human Condition and Water

Francisco Javier Galán Pérez

Climate Through the Eyes of Culture and Science

Awareness of climate change has evolved over centuries influenced by various factors, including scientific discoveries, technological advancements, and sociopolitical developments. We truly mean centuries given that classical civilizations, such as the Greeks and Romans, documented changes in weather patterns and climate variability (Harper, 2017), laying the groundwork for the earliest observations of natural phenomena. These observations were often intertwined with cultural beliefs and mythologies, shaping the traditional knowledge systems of Indigenous cultures based on their understanding of natural cycles and environmental change.

Thus, it is known that the first written lawsuit found in Aragon is the so-called Botorrita Bronze, also known as "Bronze II" of the four found so far. Bronze II, also known as *Tabula Contrebiensis*, holds historical

F. J. Galán Pérez (✉)
Predepartmental Unit of Fine Arts, University of Zaragoza, Zaragoza, Spain

OAAEP Research Group, Funded by the Government of Aragón with FEDER Funds, Zaragoza, Spain
e-mail: jgalan@unizar.es

A. M. Biedermann, F. J. Galán Pérez (eds.), *Interconnections between Culture and Sustainability*,
https://doi.org/10.1007/978-3-032-21061-6_11

significance as an archaeological artifact discovered, as we say, in late 1979 (Fatás Cabeza, 1981). This small 41 x 20 cm inscription is dated May 15, 87 BC, written in Latin, and contains a legal document. In its twenty lines, it recounts a dispute between the inhabitants of Salduie (present-day Zaragoza, Spain) and Alaun (present-day Alagón, province of Zaragoza) regarding a water channeling project initiated by the former. The people of Alaun felt aggrieved by this effort, which led to a legal dispute between the two communities.

To resolve the dispute, both parties agreed to submit the matter to neutral arbitrators, namely, the magistrates of Contrebia Belaisca (near the present-day town of Botorrita, Zaragoza province). Ultimately, these arbitrators ruled in favor of the people of Alaun, a decision made with the prior knowledge and approval of the Roman proconsul, Gaius Valerius Flaccus. This legal document is particularly noteworthy as it marks the first documented complaint in the Iberian Peninsula, offering valuable information on the legal and administrative processes of the Roman era and shedding light on the sociopolitical dynamics of the region's ancient communities. We can therefore affirm that in the territories of present-day Aragon, water has been a particular issue, as we know from the Botorrita Bronze, that has been a source of conflict and dispute, and therefore of progress and understanding.

Already during the Scientific Revolution of the seventeenth century, significant advances in climate science occurred, with studies in meteorology and climatology leading to a more systematic understanding of the Earth's atmosphere and climate dynamics. Pioneers such as Tycho Brahe, Johannes Kepler, and Galileo Galilei had contributed to the growing body of knowledge about the Earth's climate system (Gribbin, 2002), and their work would lay the groundwork for scientists to explore the intricate mechanisms governing climate processes. And already during the Industrial Revolution of the eighteenth and nineteenth centuries, rapid industrialization, urbanization, and increased consumption of fossil fuels occurred, laying the groundwork for modern concerns about climate change. Scientists such as John Tyndall and Svante Arrhenius began to recognize the potential impact of human activities on the climate during this period (Fleming, 1998). Arrhenius's groundbreaking work on the greenhouse effect in 1896 provided a theoretical basis for understanding how certain gases, such as carbon dioxide (CO_2), could influence the Earth's temperature (Quintero Plaza, 2023).

Throughout the twentieth century, advances in climate science continued, accompanied by a growing concern about environmental degradation and its potential consequences. The second half of the twentieth century witnessed heightened international awareness of environmental issues, leading to debates about environmental conservation and sustainability. The formation of organizations such as the United Nations Environment Programme (UNEP) and the convening of the first United Nations Conference on the Human Environment in 1972 reflected a growing recognition of the need for global cooperation to address environmental challenges, including climate change (Singh, 2022).

At the end of the twentieth century, scientific evidence pointing to human-induced climate change became increasingly undeniable. Rising global temperatures, melting ice caps, and changing weather patterns underscored the urgency of addressing climate change on a global scale. The establishment of the Intergovernmental Panel on Climate Change (IPCC) in 1988 provided a platform for synthesizing climate science research and assessing the impacts of climate change. International agreements such as the 1992 United Nations Framework Convention on Climate Change (UNFCCC) and the 1997 Kyoto Protocol marked important milestones in global efforts to address climate change, highlighting the growing recognition of climate change as a pressing global challenge (Breidenich et al., 1998).

Expo 2008 in Zaragoza

The 2008 Zaragoza International Exposition, held in the Aragonese city of Zaragoza (Spain), was a major global event that brought together policymakers, experts, and other stakeholders to address pressing environmental challenges (Corellano & Fraga, 2006; Corellano & Porta, 2019). The exposition's theme, "Water and Sustainable Development," reflected the growing recognition of water as a critical resource and the need for sustainable management practices to ensure its availability for future generations (Corellano & Fraga, 2006). The exposition aimed to foster dialogue and collaboration on issues related to water security, environmental sustainability, and socioeconomic development. The choice of water as a central theme was strategic, given its fundamental role in sustaining life, supporting ecosystems, and driving economic activities. Water scarcity, pollution, and unequal distribution have become major concerns worldwide, exacerbating challenges related to poverty, food security, and public health. By focusing on water and sustainable development, the exhibition sought to raise awareness of the importance of water conservation, efficiency, and equitable access as key components of sustainable development strategies (Lecardane, 2013).

The exhibition provided a platform to showcase innovative technologies, best practices, and policy solutions aimed at addressing water-related challenges. Through exhibits, conferences, statements, and publications, participants had the opportunity to explore a wide range of topics, including water management, sanitation, ecosystem conservation, and/or climate resilience. Expo 2008 also featured contributions from international organizations, governments, NGOs, and private sector entities, highlighting the collaborative nature of efforts to promote water security and sustainable development.

In addition to showcasing solutions, the Expo facilitated discussions on policy frameworks, governance structures, or financing mechanisms to support water-related initiatives. It served as a forum for exchanging knowledge, sharing experiences, forging partnerships among diverse stakeholders. By bringing together experts and practitioners from around the world, the Expo aimed to catalyze action and inspire collective efforts to address water challenges within the broader context of sustainable development. Thus, the 2008 Zaragoza International Exposition played a crucial role in raising awareness of the importance of water and sustainable development on a global scale. By highlighting the interconnections between water, the environment and world's socioeconomic development, the EXPO underscored the need for integrated approaches to addressing water challenges as well as promoting sustainable development goals (Ferren, 2008).

And within the event grounds, and beyond the more than 100 countries that attended, there were large-scale exhibitions comprised of the collaborative work carried out by Sociedad Expoagua S.A. (The public entity comprised of the Government of Spain, the Government of Aragon and the Zaragoza City Council), supranational entities, universities and design and exhibition design companies. Visitors could visit the so-called themed plazas that housed the exhibitions "OIKOS, Water and Energy," "Extreme Water," "Thirst," "Water Cities," and "Shared Water." Newly constructed buildings would house, for example, the "Water, a Unique Resource" exhibition in the Bridge Pavilion (designed by architect Zaha Hadid) and the "Water for Life" exhibition in the Water Tower. It is this last building and the exhibition it housed that we will focus our attention on.

The Water Tower and the "Water for Life" Exhibition

This building is located in the central part of the Ranillas meander in Zaragoza, and during the event, it served as the northern entrance to the Expo site from the Metropolitan Park (also built for the Expo). It is the most visible structure at the International Exposition and is a highly recognizable element, both for its volume and its distinctive teardrop shape, supported by a large plinth. The interior layout of the building is predominantly hollow, with the inner part of the facade used to allow visitors to access the exhibition content via ramps arranged in a spiral along its length.

The Water Tower's exhibition content was "Water for Life," an open exhibition without a linear narrative. Instead, visitors could construct their own narrative as they explored the different exhibition areas. These areas offered an experience through the five senses, emphasizing design and new audiovisual technologies, while also housing the scientific content of the exhibition (Del Tura Bovet et al., 2008). The exhibition was divided into two areas, following the architecture of the building itself: the plinth and the tower (Olga Subirós Studio, n.d.) (Table 11.1).

- The plinth, a large one-story hall, housed the area called "The Nature of Water," divided into the following themes: "Water Is," "Planet Water," "Blue Landscapes," "We Are Water," "The Magic of Water," "Water Cycles," and in its central space, a large sculpture entitled "Rain."

Table 11.1 Technical sheet of the different authors of the "Water for Life" exhibition project

Exhibition title: "Water for life"	
Promoter	Expoagua Zaragoza 2008 S.A.
Scientific directors	María del Tura Bovet Emma Pérez-Chacón
Scientific direction collaborators	Yolanda Arencibia, Isabel Bovet, Mario Gil, Luis Alberto Longares, Carme Pardo
Exhibition design	PROGRAM COLLECTIVE: Mona Kim, Todd palmer, Olga Subirós, Simon Taylor
Design direction	Mona Kim, Todd palmer, Olga Subirós, Simon Taylor
"Splash" design	IKONICarts: Pere Gifre, Daniel Molina, Ricardo Puertas, Alex Gamell, Paco Cavero, Albert blanch

- The tower's "Water in Mind" area was dominated by a vertical view and a smaller, exceptionally large floor plan, with a memorable highlight: the sculpture Splash. The tower is dedicated to the cultural dimension of water and was structured into the following sections: "Water for Life," "Water in Mind," "Splash," and "Cloud."

The exhibition themes were then arranged as follows:

- "Water is": Access to the exhibition was through a curtain of water accompanied by sounds of very different textures, always related to fresh water, stagnant water, rivers, lake shores, the splashing of water against wood, etc. The water curtain was bathed in two shades of blue, where the darker, more uniform color marked the fixed sections, while a lighter shade indicated the retractable, mobile part of the curtain, which would change color each time it was opened. After crossing the water curtain, access was gained to the rest of the plinth.
- "Blue Landscapes": A series of cylinders placed at different heights formed a horizontal video, unfolding in the shape of an arc, showing the main landscapes of the Earth. Each of the cylinders showed how water creates, defines, and sustains the different ecosystems existing on our planet. These videos demonstrated the creative power of water in each landscape and allowed us to capture the subtlety with which they develop as water molds and shapes the landscape. Although some of these changes are natural and cyclical, certain human actions pose a threat to the balance of the planet: mass tourism, forest fires or extensive agriculture compete with the traditional lifestyles on which human existence has been based for centuries. Blue Landscapes showed the delicate balance between the environment, water and humanity.
- "We Are Water": This was a tunnel-shaped installation with information on the exterior about how much water we drink daily, explaining why our bodies excrete water and exploring the importance of water in different organs. Above all, it highlighted the importance of water in the bodily processes that allow us to live: from cells to blood, water is an essential element for the functioning of our bodies. Upon entering the tunnel, visitors crossed through different multisensory spaces: warm and dry, cold and dry, warm and humid. There, visitors could participate both physically and intellectually in the different

environments and perceive how their sensations materialize into biological responses related to water.

- "The Magic of Water": This section was structured around a series of experiments that explained in a simple way some of the most relevant physical and chemical properties of water, highlighting those that allow us to understand the importance of this element for life. Each experiment was presented on two scales: a macroscopic one (planet, landscapes); and a detailed one, combining examples from everyday life with others of a microscopic or molecular nature (ions, molecules, or aspects that can only be observed through a microscope).
- "Planet Water": A multi-screen section displayed a rapid-fire timeline, showing statistics related to the planet's origin and the presence of water on Earth. The story spanned a period that began billions of years ago and continued into the present day, reliving moments that marked the process by which water, a unique chemical element present in stars, exploded within a desolate planet, transforming it into a planet of water and, consequently, a planet of life.
- "Water Cycles": This section invites visitors to reflect on the comparative nature of water in the human body. A parallel is established between the act of drinking a beverage and the way rain irrigates the earth. The location of "Water Cycles" in the center of the plinth, a monumental Rain Fountain (a symbol of the water cycle) intensifies the significance of this installation. "Water Cycles" suggested that our lives and those of the planet are inseparable.
- "Rain Fountain": In the center of the plinth was a fountain made up of two monumental elements: a cistern and the rain falling on it, representing the meeting of the waters of the earth and the waters of the sky. Rain was a sculpture made up of several transparent cylinders representing enormous columns of rain, whose fall from the top of the plinth's atrium had been suspended.
- "Collapse": The "Water for Life" exhibit showed the different ways in which water shapes life on Earth. To effectively capture this message, every fourteen minutes the exhibition underwent a spectacular transformation and flooded. This was the "Collapse". As the "Collapse" began, the blue landscapes filled with clouds and took on a dark blue hue. The rain grew in intensity until it became a deafening torrent, until it suddenly ceased to be heard. Meanwhile, the whole exhibition was bathed in blue. Visitors were also bathed in the color blue. The message in this case was clear: we are all water.

- "Water for Life": Upon reaching the Tower level, visitors encountered large screens showing a film whose plot highlighted the importance of water for human life. The plot unfolded like a kind of visual puzzle, as the visitor left the plinth and ascended the ramps, completing the story. The central theme of the film was how, through water, one life intersects with another and shares feelings, experiences, thoughts, and choices (Olga Subirós Studio, n.d.). In that story, water constituted the framework for the action, bringing about changes and transitions and forming part of an individual's rituals and habits. But, above all, water, in its various forms (in the bath, in a shower, in a religious vessel), evoked memories through which the visitor recognized human life and culture.
- "Water in Mind": From this point on, there were two options: either leave the exhibition through a designated exit or continue the ascent to see the remaining sections: "Water in Mind," "Splash," and, at the end of the tour, cross the cloud crowning the tower. Thus, while ascending the ramp, the visitor encountered seven landings, each with an installation. Each of these seven installations revealed some aspect of the cultural value of water through seven fundamental moments in human life.
- "Splash": As the visitor ascended the ramp to reach and pass through the cloud, they encountered the suspended sculpture "Splash," which we will focus on in the next point and is also the axis of the present article. "Splash" represents the dynamic moment when the forces acting on a droplet split it into an infinite number of smaller drops that splash. The design and execution of "Splash" combine art, media, science, engineering, and technology. The form of this work was developed using advanced modeling software that predicts and recreates the complex geometry of fluid dynamics. "Splash" represents water as a force, energy, and the possibility of life, even at a scale as small as that of an insignificant droplet. As the visitor ascended the ramps toward the cloud, they circled around the sculpture from different heights and angles.
- "Nube Bar": Located on the top floor of the Water Tower, the Nube Bar (Nube stands for "cloud" in Spanish language) was the culmination of the "Water for Life" exhibition, at the highest point in the building. It was a semi-open floor of the Water Tower, conceived and designed as a viewing platform with excellent panoramic views of the International Exhibition grounds and the city of Zaragoza.

A Suspended Sculpture

Technically speaking, “Splash” is a sculpture measuring approximately 23 meters high and 5 to 9 meters wide, made up of 134 pieces reproducing a splash of water, with a chrome finish on its surface. It is suspended by steel cables anchored to an intermediate floor of the building, thus remaining suspended within the 40-meter-high interior void at the top of the Water Tower (Asociación Legado Expo, 2020). The body of each piece is made of WRF polyurethane foam, covered with expanded polystyrene, and finished with chrome paint over fiberglass. The pieces are reinforced with solid iron profiles and hollow aluminum profiles, and all have one or more steel cables for subsequent hoisting onto the floor (Fig. 11.1 and Fig. 11.2).

Fig. 11.1 The Splash sculpture seen from below. Visitors walk along the circular path of the interior facade. *Splash* (views from the base/interior of the Water Tower). (Photograph by Zaragoza al natural, CCBY2.0)

Fig. 11.2 The Splash sculpture seen from above. Visitors walk along the circular path of the interior facade and through the hall. *Splash* (detail of the suspended splash). (Photograph by Zaragoza al natural, CCBY-SA2.0)

End of the Exhibition, Dismantling, Repair and Assembly

At the end of the exhibition, the exhibition was dismantled. However, with respect to the "Splash" sculpture, there were three options:

- The Expoagua Society could decide to keep it and have it permanently installed inside the building.
- The Expoagua Society could decide to dismantle it and send it to a landfill in accordance with current regulations.
- The Expoagua Society could decide to dismantle and package the sculpture and store it in a warehouse for later use.

Unfortunately, because the planned use for the Water Tower, although unfinished, vague, and basically nonexistent, did include the sculpture's defeasance, it was decided to dismantle it and precariously conserve it in the Zaragoza City Hall warehouses. Furthermore, removing it from the building required a larger disassembly than planned, given that the gaps created to pass inside the sculpture's various elements had been subsequently reduced with the addition of various building installations.

Shortly after "Splash," already dismantled and packed, was removed from its original location, the entity that demanded its removal to receive the building, the Caja de Ahorros de la Inmaculada de Aragón (a local Bank), would disappear following a merger with two other savings banks in late 2010 (which resulted in a new savings bank called Caja3), and that supra-entity would finally be acquired by Ibercaja Banco three years later. This raised hopes of recovering the sculpture for relocation to the same imposing hall of the Water Tower, and the first steps were taken in the summer of 2013, culminating in its reintroduction to its original location around mid-2014 (Asociación Legado Expo, 2020), six years after the Zaragoza International Expo.

Thus, to this day, although the Water Tower remains closed to the public except for tours organized by the Expo Legacy Association and a very limited and infrequent number of events held in the building since 2009, no one has access to the Water Tower and, therefore, to the sculpture. Although the reasons for this underutilization and its very possible solution are worth asking and subsequently answering, it is beyond the scope of this article to go into detail.

The Sculpture "Splash" Through Time

The reason for the sculpture's enduring character lies in a harmonious convergence of traditional artistic expression and cutting-edge technological innovation. Conceived by artist Pere Gifre, this dynamic work of art combines the aesthetics of sculpture with the precision of digital design. This is because a central element of the creation of the "Splash" sculpture is the revolutionary technique of digital fabrication. By harnessing the power of computer-aided design software and advanced manufacturing and machining technologies, the exhibition's design team was able to transform their artistic vision into a tangible reality with unprecedented precision and complexity.

As a symbol of twenty-first-century art and technology, the sculpture "Splash" pays homage to the evolution of artistic expression in the digital age. Beyond its extra-dimensional volumes, "Splash," when visible to the public, quickly achieved iconic status at the Zaragoza International Exposition and also within Zaragoza's cultural landscape, serving as a symbol of artistic innovation and technological progress. Its prominent location within the Water Tower not only enhances the city's architectural skyline but also invites visitors to contemplate the intersection of art, technology, and human creativity. In June 2008, the New York Times headlined "It's Zaragoza's Turn to Make a Splash" a clear play on the city's potential for internationalization through its heritage and the efforts made at Expo 2008. Today, the story of this unique piece is beginning to become a hidden secret of the city that needs to be properly publicized. If anything characterizes the city of Zaragoza, it is its vast cultural heritage, which, like any other place in the world, requires explanation, knowledge, and care. As we explained at the beginning of this article, the city is a palimpsest where the Iberian, Roman, Visigothic, Muslim, Jewish, and Christian legacies converge in all their expressions throughout the ages (Romanesque, Gothic, Mudejar, Baroque, Renaissance, etc.), as well as the artistic production closest to our time, such as cinema, photography, the avant-garde, and art during the transition to democracy. Thus, to our particular legacy, for the moment almost hidden, we have added the sculpture "Splash" as our own particular icon of the twenty-first century.

Conclusions

Talking about sustainability in the cultural sphere implies going beyond the efficient management of resources or the reuse of materials. It also entails considering how the symbols, narratives, and spaces that shape collective identity are preserved, interpreted, and activated. "Splash" offers a paradigmatic example of how a work of art can embody the tensions between sustainability discourses and institutional practices that, rather than sustaining, suspend or interrupt that same symbolic continuity. Thus, cultural sustainability, in its deepest dimension, implies keeping alive the narratives that help a society reflect on itself. Because "Splash" was not only a large-scale sculptural piece, but a semiotic artifact that captured (in form, materiality, and placement) a vision of the future in which water is not just a resource, but a structuring principle of life, culture, and ecological awareness. As an installation, it was a manifestation of the culture of sustainability; as an object, it carried environmental memory. None of that mattered (nothing mattered, we dare to say) and the abrupt dismantling of "Splash" after the exhibition closed and its subsequent concealment in municipal warehouses was not simply a logistical matter but constituted a suspension, also symbolic, of the narrative the sculpture embodied. In cultural terms, dismantling "Splash" meant, in a way, deactivating the collective message the event had sought to consolidate: that sustainability was not just an environmental or technological issue, but also an ethical and aesthetic commitment to the future. This interrupted act, which affected a central element of the Expo's visual discourse, illustrates how institutions can suspend the values they proclaim, dismantling the cultural meaning of sustainability.

The subsequent reinstallation of Splash, although materially valuable, failed to fully reverse this suspension of meaning. The fact that the sculpture returns to its original space, but remains invisible to the public due to the Water Tower's prolonged closure, adds an additional layer of complexity: what is sustainable is no longer just what is preserved, but also what is activated. Cultural sustainability requires a continuity of presence, experience, and interaction. Without access, without narrative, without mediation, the work becomes an echo of the past rather than a promise of the future. Yes, another one.

This case allows us to reflect on the way recent cultural legacies (and especially those linked to ephemeral events such as international exhibitions) are managed in contexts of political, economic or institutional transition. Instead of being integrated into the living fabric of the city and its citizens, they often fall into a sort of heritage limbo, suspended between memory and oblivion. Thus, cultural sustainability is challenged by an administrative inertia that postpones the active care of contemporary symbols while simultaneously exalting already canonized historical heritages.

Ultimately, "Splash" is more than a suspended sculpture: it's a metaphor for the ambiguities of our relationship with the future. It represents the potential energy of a culture that aspires to be sustainable, but whose realization is often conditioned by factors beyond art and critical thought. Remembering and reactivating "Splash" is also an act of symbolic re-inscription: a way of reaffirming that sustainability, as a culture, only exists when it is sustained over time and in collective experience.

References

Asociación Legado Expo. (2020, marzo 5). *Splash, una obra de Pere Gifre*. Asociación Legado Expo Zaragoza 2008—Legado Expo. https://www.legadoexpozaragoza.com/legadoexpo/noticias/splash-una-obra-de-pere-gifre/

Breidenich, C., Magraw, D., Rowley, A., & Rubin, J. W. (1998). The Kyoto Protocol to the United Nations Framework Convention on Climate Change.

American Journal of International Law, 92(2), 315–331. https://doi.org/10.2307/2998044

Corellano, F. P., & Fraga, F. J. M. (2006). El proyecto Expo 2008: arquitectura, ciudad, arte público. *Artigrama, 21*, 197–213. https://doi.org/10.26754/ojs_artigrama/artigrama.2006218136

Corellano, F. P., & Porta, M. P. S. (2019). Grandes eventos, huellas del futuro. Las riberas del Ebro y Expo Zaragoza 2008. *Zarch: Journal Of Interdisciplinary Studies In Architecture And Urbanism, 13*, 62–75. https://doi.org/10.26754/ojs_zarch/zarch.2019133912

Fatás Cabeza, G. (1981). El Bronce de Contrebia Belaísca. In *Cuadernos de trabajos de la Escuela Española de Historia y Arqueología en Roma*. Arqueología (EEHAR). http://hdl.handle.net/10261/160059

Ferren, A. (2008, junio 12). Journeys: Expo 2008 in Zaragoza, Spain. *The New York Times.* https://www.nytimes.com/2008/06/15/travel/15journeys.html

Fleming, J. R. (1998). *Historical perspectives on climate change.* Oxford University Press.

Gribbin, J. (2002). *Science: A history, 1543–2001.* Allen Lane.

Harper, K. (2017). *The fate of Rome: Climate, disease, and the end of an empire.* Princeton University Press. https://doi.org/10.1515/9781400888915

Lecardane, R. (2013). Il territorio dell'acqua: Saragozza dopo l'Expo 2008. In *Agathón /RCAPIA PhD Journal* (pp. 41–46). OffsetStudio.

Quintero Plaza, D. (2023). El efecto invernadero: descripción y descubrimiento. Calendario meteorológico, pp. 243–250. Agencia Estatal de Meteorología. https://doi.org/10.31978/666-23-007-X

SA, E. Z. 2., Del Tura Bovet, M., Pérez-Chacón, E., & De Teresa, E. (2008). *Agua, vida.*

Singh, K. (2022). Looking through Indira Gandhi's vision: Some recollections. *Environmental Policy and Law, 52*(3–4), 1–5. https://doi.org/10.3233/EPL-219037

Water for Life—Olga Subirós Studio. (n.d.). Olga Subirós Studio. https://olgasubiros.com/project/water-life-exhibition-expo-zaragoza-2008/

CHAPTER 12

Strategies for Developing Sustainability and Environmental Education in Collaborative Art Practices: The Case of Trash + Fantastic Beasts

José Prieto Martin and Fabiane Cristina Silva Dos Santos

Introduction

Sustainable development has ceased to be a long-term objective and has become a non-negotiable condition. The term originated in 1987 when the World Commission on Environment and Development published "Our Common Future" (Brundtland Report). It defined it as meeting "the needs of the present generation without compromising the ability of future generations to meet their own needs" (Brundtland Report, 1987, p. 23). However, for the time being, we are unable to resolve the major environmental crisis and evolve toward sustainable development, given that economic interests prevail over environmental interests. For this

J. Prieto Martin (✉) • F. C. Silva Dos Santos
Predepartmental Unit of Fine Arts, University of Zaragoza, Zaragoza, Spain

OAAEP Group, Funded by the Government of Aragón with FEDER Funds, Zaragoza, Spain
e-mail: perancho@unizar.es; fsilva@unizar.es

A. M. Biedermann, F. J. Galán Pérez (eds.), *Interconnections between Culture and Sustainability*,
https://doi.org/10.1007/978-3-032-21061-6_12

reason, education must foster responsible behavior by individuals and communities regarding the protection and improvement of the environment.

In the artistic field, there are also strategies for developing sustainability. The Trash Art movement uses recycled materials and sustainable techniques to create its works, which minimize environmental impact, promote social and ecological awareness, and encourage reflection on excessive consumption. Its working methods and artistic representations are the result of incorporating waste materials generated by our consumer society as an artistic resource, without hiding their origin. These materials, when reused, acquire cultural value. Strategies for developing sustainability and environmental education in collaborative artistic practices. The specific case of Trash + Fantastic Animals is a continuation and extension of the shared artistic, educational, and authorship project, Trash + Bestiary. It promotes reuse or recycling, individually, collectively, and collaboratively, among professors and students of the Fine Arts degree at the University of Zaragoza (Teruel Campus) and students of the Fine Arts degree at the Faculty of Fine Arts of the University of Porto. It is designed to encourage and produce collaborative gestures that provoke intersections between Teruel's Mudejar heritage, contemporary art (Trash Art), recycling, and sustainability. It addresses consumer habits and incorporates waste generated in the immediate surroundings as both an educational and artistic resource to reinterpret the bestiary on the ceiling of Teruel Cathedral.

Collaborative Artistic Practices, Ecological Awareness, and Heritage Value

During the nineteenth and twentieth centuries, artists took on an important role within artistic movements as agents of social change. Following avant-garde discussions surrounding social and political issues, artists began to actively engage with space and its social issues. The Dada and Surrealist movements introduced collective collaborations into their work as a way of subverting the status of the individual artist, a challenge that began to break with the canons of art.

The artist begins to seek new strategies for his production, and in this search, he turns creation into an instrument of penetration and social action that questions common sense, thereby attempting to impact a counter-hegemonic discourse. We take as a precedent Bertolt Brecht's theory and practice of theater, which articulated art, theatrical work, and social praxis, resulting in the production of mass art that contributed to unmasking the ideology of the dominant class and attempted to raise awareness about social inequality. But for Rancière, the vision of the spectacle does not simply end in intellectual understanding or awareness of the world. From there, the viewer would have to decide to act, thus moving to another sphere: from a sensitive vision of the world to one that defines other visions of tolerance and intolerance, abilities and inabilities. And starting from that perspective, the viewer would end up recognizing the possibility of making an active and not only emancipatory interpretation, in which the political value of art is not found in the content, but in the forms of the process that makes it visible. Emancipation from art would involve the recognition that the production of the work does not belong to its creator.

In the 1960s, social movements such as environmentalism, feminism, and identity politics began to activate collaborative processes, working on their cultural processes alongside a specific community. A collaborative art movement or community art emerged, a highly developed artistic movement in the United States, the United Kingdom, and some Nordic countries. This movement marked a space between the political activism of the 1970s and 1980s and popular arts. Popular education and cultural workers sought to generate a direct relationship with the context, betting on participatory art and an active conception (Santos, 2021, pp. 27–28). From the 1980s onward, the relationship between art and politics became more pronounced, with a multiplication of artistic initiatives through which artists sought to promote the participation of members of the public in projects such as experimentation with forms of socialization and occupation of local spaces.

We are therefore beginning to see new paradigms, both aesthetic and cultural, according to which art is given a social function: public space becomes an intervened place that is directly related to art and its local context, and the viewer takes on a fundamental role in the creative process. All this unconventional change in the artistic field and the growing interest in public art contribute to the beginning of a new era in artistic production, which now embraces a new way of conceptualizing and giving visibility to artistic practices that are directly related to changes in new forms of political activism. Through the latter, civil society is taking a more participatory stance within a process of consolidating citizenship through emancipation.

Collaborative artistic practices are emerging within the expanded field of contemporary visual arts, with the active participation of the community in a process of transversality with other disciplines. Nicolas Bourriaud called it relational art in the 1990s. This shifts the focus from the work as an object to a process of experience that opens up to social involvement in the political, cultural, and ecological fields. Art becomes a space for mediation and creation, articulating reflections on issues of identity, memory, territory, and ecology. Collaborative art thus emerges, following the lines of political art. The artistic proposals were found within collectives with certain types of social problems, and the aim was to involve citizens in resolving the conflicts faced by these groups, thus making the problems visible, fostering integration, and reinforcing the self-esteem of its members.

Collaborative art, therefore, seeks to rethink the role of the artist, the audience, and the work, questioning the traditional conception of art. The artist is placed on an equal footing with the viewer, and the viewer ceases to be a passive recipient and becomes actively integrated into the process of developing the work. The work also receives a new perception, which emerges in relation to its context, and the process gains significance from the collective work of its actors and their actions: the resulting work ceases to be a final outcome and becomes direct evidence that presents a result derived from the multiplicity of individual creative actions. In this process of developing the proposal, for its purpose to be achieved, it is important that everyone participate in the development of the project; thus, a methodology is generated so that, in general, all participants interact with each other, fostering the integration of different groups that coexist in a given social context.

> *Art is supposed to be political because it shows the stigmas of domination, or because it ridicules the reigning icons, or even because it leaves its places to become a social practice, etc.* (Rancière, 2010)

In the context of the ecological crisis, collaborative art is gaining particular relevance, as it becomes a facilitator for environmental education dynamics, in a process of eco-social regeneration that promotes the strengthening of biocultural heritage. In this context, contemporary artistic practices with an ecological conscience emerge, among which we highlight Trash Art, where the use of recycled materials, industrial waste, or urban waste is the basis for artistic creation. The antecedents of this artistic movement can be traced back to the artistic or historical avant-garde movements of the early twentieth century, in creators who ventured to incorporate all kinds of materials and objects into their works, such as George Braque and Pablo Picasso, who in 1912 with his Papiers collés made a capital innovation by adding real objects to their paintings, destroying the homogeneity of painting. Or like Umberto Boccioni, who stated: "*We reject the exclusive use of a single material … we affirm that up to twenty different materials can be used in a work to achieve a plastic emotion … glass, wood, cardboard, iron, cement, horsehair, feathers, fabric, mirrors, electric light, etc.*" (Ruhrberg et al., 2005, p. 434). This paved the way for material culture and the world of objects, which would be fully developed by Marcel Duchamp, among others, with his Ready-mades, transforming mass-produced objects into works of art; and Kurt Schwitters, whose work focused on collage created a new concept of artwork with his Merzbau, which began as studio sculpture and evolved into a "votive chapel" filled with assembled found objects. This predisposition to recycling waste materials found its followers in Italy with Arte Povera, which used humble and poor, generally non-industrial, materials. Their transformation was valued primarily, as they deteriorated, transforming the work. This work takes on new force in the current context of environmental crisis and becomes a tool of social critique against capitalist hyper-productivity and disposable consumption models. It promotes reflection on the life cycle of objects, waste management, planned obsolescence, and pollution, in a process of critical pedagogy that we can link to the thought of Paulo Freire, who argued that the transformation of the world begins with the recognition of the immediate material environment. In this case, waste becomes not only an aesthetic medium, but also a political and epistemological one, highlighting the circuits of exclusion, accumulation, and inequality that structure our economies and landscapes.

The Trash + Fantastic Beasts project, based on collaborative practice and a process of collective pedagogy, working with Fine Arts students, uses valueless materials—garbage, waste, and other urban debris—found in today's throwaway consumer culture, transforming them into sculptures. This project addresses art, education, and activism transversally, promoting a wake-up call about consumption and our relationship with the environment. It proposes a co-creation where project participants contribute their ideas, creating the works together.

> *Collaborative practices carry out a reconceptualization of the functions that art and culture can develop in public spheres, once their action is linked to the rhythms, tensions, and conflicts that currently emerge. Their contextual nature entails assuming them as a critical and political activity, which seeks to question both the places and roles distributed within the field of culture itself, as well as the ways of life and conditions of other frameworks from which the previous ones are no longer so separated.* (Collados & Rodrigo, 2015: pp. 57–72)

Trash + Fantastic Animals proposes a collective, community-based work based on art, emphasizing the local context in a dialogue with the historical heritage of the city of Teruel and Mudejar art, in a value as an approximation of the heritage context to the university educational community and local schools, within a dialogic aesthetic emphasizing the local space as a place of production and generation of relationships, highlighting social relationships with the environment, generating spaces for interaction, resulting in a participatory work.

In dialogue with the *Trash + Fantastic Animals* project, we find collaborative, pedagogical, and territorial artistic projects, involving local communities in processes of collection, classification, and artistic transformation of materials considered "garbage." For example, Brazilian artist Vik Muniz worked with a collective of garbage collectors to create large-scale photographs. The project involved taking photos of people collecting in landfills and, from these photos, reconstructing waste collages and then photographing them. He also works with expired food. Mexican artist Betsabeé Romero has worked with Indigenous communities and urban recyclers to transform used tires into monumental installations. Her work explores the links between mobility, waste, and popular culture, using rubber as a symbol of cultural resistance to disposable modernity.

In this sense, the collaborative artistic process is situated within what is called eco-social aesthetics in which art is articulated in a process of ecological transformation and social justice (Demos, 2020), which generate new forms of expanded heritage, which are not focused on objects of value, but on the search for the collective capacity to build memory and future from the margins, giving a new meaning to waste and seeking awareness in relation to social problems and the environmental crisis.

The Particular Case of Trash + Fantastic Beasts

This research is an innovative proposal that, within a theoretical framework, links two concepts that work together: Art Education and Environmental Education, to raise awareness of contemporary environmental issues, seek ways to address them, and reflect on some Sustainable Development Goals of the 2030 Agenda.

Art Education, on the one hand, is centered on the Trash Art artistic movement, which primarily practices the technique of assemblages, creating imaginative configurations and proposing alternative ways of working while caring for the environment in a sustainable manner. There are many contemporary creators within this movement who have proven that beauty or social criticism can emerge from a pile of trash, showing us that creativity and ingenuity can be fueled even by the waste associated with rampant hyper-consumption, in a world where millions of people still die of hunger. Trash Art gained significant media attention following the exhibition *"Apocalypse—Beauty and Horror in Contemporary Art,"*[1] held from September 9 to December 15, 2000, at the Royal Academy of Arts in London.

> *This autumn the Royal Academy will present Apocalypse, a major exhibition of international contemporary art which will reflect aspects of the richness, beauty, horror, complexity and diversity of the world today. Concentrating on themes inspired by the arrival of the 21st century, the exhibition will feature some thirteen installations, paintings, sculptures and video works, the majority of which have never been seen in public before. Apocalypse is a story of extremes. This exhibition is a contemporary, secular interpretation of the biblical story of St John the Divine which contains elements ranging from the horrors of genocide to the beauties of Utopia. These elements are represented in the exhibition which aims to demonstrate the possibilities art holds in the world in which we live. The exhibition encompasses all media from painting and sculpture to multimedia works using the latest technology. Anticipated works in the show range from a monumental new piece by the Chapman brothers—Hell—depicting the horrors of 20th-century genocide, to Mariko Mori's highly spiritual 21st-century Dream Temper. Powerful new paintings by Luc Tuymans will be included as will a video installation by cult pop video maker Chris Cunningham, whose*

[1] The artists participating in the exhibition were: Darren Almond, Maurizio Cattelan, Jake and Dinos Chapman, Chris Cunningham, Angus Fairhurst, Mike Kelley, Jeff Koons, Mariko Mori, Tim Noble and Sue Webster, Richard Prince, Gregor Schneider, Wolfgang Tillmans, and Luc Tuymans.

work has never previously been exhibited in a gallery, accompanied by a soundtrack by Richard James, the Aphex Twin. (Royal Academy of the Arts, London, 2000)

Trash art continues to thrive today, generating a significant impact both in the artistic sphere and in raising environmental awareness about sustainability. There are artists/artivists in this movement who contribute to environmental awareness campaigns, cleaning beaches, oceans, and rivers, such as Handiwirman Saputra (Indonesia), who has developed an "aesthetic art of waste, of used objects" in collages he constructs with various materials found in waste disposal containers or wood from riverbanks. In 2007, American Mark Dion carefully classified trash found in the Thames in his installation *Fieldwork 04* at the PRAE (in collaboration with the Natural History Museum in London). German HA Schult built his *Save the Beach Hotel* in 2010 with trash collected from beaches in southern Europe. STUDIOKCA (formed by Jason Klimoski and Lesley Chang), in their work *Cloud Room One New York Minute* (2014), created a cloud with the number of plastic bottles that are thrown away, more or less, in one minute in the United States.

Other artists photographically document the mark that garbage leaves on our planet in landfills and oceans, such as the American Chris Jordan, from 2003 to 2005, and in his project *Albatross* (2017), he revealed the garbage that can be stored in the entrails of a dead bird; and, the Brazilian Vick Muniz, who creates large-scale images with objects that he collects with the collective of garbage collectors in landfills and subsequently photographs them. Others collect, manipulate, store, classify, and reuse garbage (from dumps, or from events in cities) to build two- and three-dimensional works, such as the Chinese Xu Bing, in *The Phoenix Project*, creates giant winged animals with different materials found in demolition dumps in urban areas of China. And, in 2001, the American Justin Gignac collected trash from New York, then packaged it in small transparent methacrylate cubes (signed, numbered, and dated) and sold it (Moszynska, 2013).

Some creators practice upcycling, or super-recycling, such as the Frenchman+Brauer, who uses discarded industrial materials to create illuminated, retro-futuristic, and peaceful robot sculptures, or the Spanish Marta Sanmamed, who reuses plastic bags and weaves them into spectacular vertical gardens.

There are also artists who, in addition to recycling, experiment with discarded materials and light. The pioneer is Dutch artist Dieter Wiegman, who has worked since the 1970s with shadows projected by piles of garbage—"light sculptures." In these, the shadows portray a different image than the piles of waste, transforming ugliness into beauty. This practice was later adopted by Tim Noble and Sue Webster, English artists associated with the post-YBA generation and influenced by punk rock.

And there are other artists who work on city streets and are more closely linked to Street Art or Postgraffiti, such as the Portuguese Bordalo II, who creates murals on buildings and walls using trash and scrap metal; they generally depict stray animals staring at us. And the Spaniard Francisco de Pájaro, who in 2009 began a super-ephemeral street art project called "*Art is Trash*" painting directly onto the trash on the streets of Barcelona shortly before it's collected.

Art Education also focuses on understanding Teruel's rich Mudejar heritage, considering that cultural heritage helps build and maintain a community's identity and memory, connecting it with its past and roots. This project-process reinterprets some animals from the Bestiary of the Roof of the Cathedral of Santa María de Mediavilla in Teruel, known by some historians as the *Sistine Chapel of Mudejar art.*

If the ceiling of the Teruel Cathedral can be described as an exceptional work of art, even if only considering the question of the uniqueness of its structure, when finally referring to its painted decoration, the piece undoubtedly deserves the qualification of "unicum," which makes it known by the nickname of "Sistine Chapel of Mudejar art" (Borrás, 1999, p. 27).

Environmental Education seeks to develop citizens who are conscious and committed to environmental care and sustainable development. Our society is dominated by a throwaway culture. This is aggravated, among other things, by the enormous population growth that consumes at a rapid pace and generates a large amount of waste that piles up daily in large cities, causing high pollution and a major global environmental crisis that impacts, among other things, climate change. It is important to consider that, beginning in the 1960s and 1970s, efforts have been made to address this problem. The United Nations Educational, Scientific and Cultural Organization (UNESCO) developed a dossier on Environmental Education that has served as the basis for the following proposals. Since then, Environmental Education has been the central theme of various international congresses and conferences, among which one can highlight:

the International Colloquium on Environmental Education (Belgrade, 1975); Intergovernmental Conference on Environmental Education (Tbilisi, 1977); International Congress on Environmental Education and Training (Moscow, 1987), which defined the concept; and the International Conference on Environment and Society: Education and Awareness for Sustainability (Thessaloniki, 1997).

Background of the Study Case and Current Status

This project-process is, on the one hand, a training activity and, on the other, an experience of creative self-expression and collaborative and participatory work. It involves teachers, university students, schoolchildren, artists, and administrators. It has been carried out in different academic contexts in Teruel from the 2018–2019 academic year to the 2023–2024 academic year.

Artists José Prieto and Vega Ruiz brought this artistic process project into the 4th grade classrooms at CEIP Las Anejas, transforming them into a different learning space than the students are used to in their daily lives. In collaboration with Sciences teacher María Játiva, they worked with the students, acting as artists interconnecting ideas. To carry out the experiment, guidelines were established, such as having the students rescue plastics from their household trash, then clean them, store them at school, and use them to create a sustainable bestiary. At the same time, the students were taken to the cathedral to analyze and gain a deeper understanding of the roof bestiary and were introduced to the Trash Art movement (Prieto & Ruiz, 2020, pp. 91–92).

Currently, the project process has expanded its scope of work (from local to global) and is being developed with members of the Ibero-American Network for Research in Art, Memory, and Environmental Sustainability [ArteM(y)SA]: professors from the Faculty of Social and Human Sciences at the University of Zaragoza: José Prieto, Fabiane Santos, and Pedro Luis Hernando (director of the Museum of Sacred Art of Teruel); professor Teresa Almeida from the Faculty of Fine Arts at the University of Porto; and visual artist Vega Ruiz. And with students from the Fine Arts degrees at the University of Zaragoza and the University of Porto. Between February 17 and 24, 2025, José Prieto completed an Erasmus teaching stay at the Faculdade de Belas Artes da Universidade do Porto to develop an artistic work inspired by Portuguese heritage, specifically tiles, which are an integral part of the architecture and culture of this city, holding an exhibition at the Faculty of Fine Arts of the University of Porto, featuring the results of the research. This new line of research emerged from this stay (Fig. 12.1).

The implementation of the first part was done online, since this communication medium favors collaborative work and develops communication and teamwork skills. The second part, the material creation, was carried out in the Fine Arts laboratories of Teruel and Porto, where several interpretations of the bestiary of the Mudejar ceiling of Teruel Cathedral were constructed using recycled material. This activity took place during the months of March, April, and May, between students from Teruel and Porto, who communicated synchronously, that is, in real time, with each side receiving (and, if necessary, processing and responding) messages

Fig. 12.1 Wild Pig. Mixed media. 90 × 60 × 160 cm. Cloister of the Museum of Sacred Art of Teruel, March 2019

instantly. They were coordinated by professors José Prieto[2] and Teresa Almeida. The third part of this research consisted of presenting the results at two conferences: "*Monstres & Chimères. Images des plafonds peints médiévaux*" at the Abbaye Médiévale, Centre cultural Les arts de lire, Lagrasse, France, on May 30 and 31, 2025 (exhibition and presentation) and "Virtual Ubiquitous and Social: Learning with ICT (*USATIC Virtual Conference*)" 2025, on June 24, 25, and 26, 2025.

Objectives and Interests

The main objective of this project-process is to raise awareness of contemporary environmental problems, seeking ways to address them, and promoting environmental values [3Rs (Reduce, Reuse, and Recycle)], to develop some of the goals of the Sustainable Development Goals of the 2030 Agenda (No. 12, No. 13, No. 14, and No. 15), to introduce participants to contemporary art (Trash Art), participatory and collaborative art, and to create works of Trash Art. Other objectives to be met are: to introduce participants (professors and students from the universities of Porto and Zaragoza) to the artistic heritage of Teruel (the Bestiary of the Mudejar Ceiling of Teruel Cathedral), to contemporary art (Trash Art), and to Environmental Education. And to generate synergies between students and faculty of Fine Arts at the University of Porto and the University of Zaragoza (Fig. 12.2).

Methodology

The methodology used to develop this project-process was, on the one hand, documentary, since the documentary review recovers the accumulated knowledge on the subject and provides an insight into its development, as Maurice Duverger explained: "Documentary research is based on the study of everything on which human beings have left their mark" (Tena Suck & Rivas-Torres, 2005, p. 49). The methodology was also active and participatory, where knowledge is acquired as the subject interacts with the environment and society. As Carlos Wohlers explained, "*active learning methodologies adapt to a model in which the primary role*

[2] Between May 19 and 23, 2025, he had another research stay at the Faculdade de Belas Artes da Universidade do Porto, as a researcher at the Aragonese Observatory of Art in the Public Sphere (OAAEP).

Fig. 12.2 Front and back covers of the exhibition catalogue: "TRASH + MEDIEVAL BESTIARY. Heritage as a Source of Inspiration," Cloister of the Museum of Sacred Art of Teruel, October 2021

falls to the student, who constructs knowledge based on guidelines, activities, or scenarios designed by the teacher" (Gálvez, 2013, p. 13).

At the same time, the working methods of this project-process are collaborative and co-authored, with an experimental approach, as students must research, experiment, manipulate, and study the creative possibilities of recycled materials so that they can be used in artistic representation. The teacher acts as a network of ideas, and students interact and reflect on the results obtained, and from there, create their own knowledge.

> *Process projects are designed as tools to help students and teachers learn and experience art, and to develop their artistic self-education. Students, like the teacher, have an active role; they are not observers or consumers of data or theories; rather, they must be provided with opportunities to develop their artistic knowledge. Projects describe this educational encounter: they identify an artistic work situation, a theme or material to be addressed, a task to be initiated…*" (Agra, 2003, p. 68)

Therefore, the proposed activity has guidelines set by the teacher, but it is the students who, through experimentation and research with the materials and their possibilities, creatively develop the entire process. As the work is collaborative, with a participatory approach and interdependence among team members, the need to trust one another arises, and the success of each is the success of all. Allan Kaprow, in his book Between Art and Life: Essays on the Happening, speaks about participatory art and indicates that it cannot be pre-planned in every detail by the artist, but must be free to emerge and grow wherever the participants want to take it (Kaprow, 2016).

Results and Impact

The results of this research have been a hybrid and sustainable bestiary constructed by assembling recycled materials (plastic, metal, cardboard, wood, cork, fabric, glass, etc.), which maintains its identity and is inspired by the bestiary on the ceiling of Teruel Cathedral. It has been shown over the past seven years in various exhibition spaces:

- Cloister of the Museum of Sacred Art of Teruel: "TRASH + BESTIARY" from March 14 to 25, 2019; and October 2021, as part of the 15th International Symposium on Mudejarism: "TRASH + MEDIEVAL BESTIARY. Heritage as a source of inspiration."
- Exhibition Hall and Hall of the Fine Arts Building of Teruel: "TRASH + BESTIARY +" (from January 9 to 31, 2020).
- Vice-Rector's Office Exhibition Hall and Teruel Campus Gardens: "TRASH ART III" (June 14–18, 2021), as part of the 15th International Symposium on Mudejarism.
- Cloister of the Museum of Sacred Art of Teruel: "TRASH + MEDIEVAL BESTIARY. Heritage as a Source of Inspiration," October 6–15, 2021.
- Vice-Rector's Office Exhibition Hall: "TRASH OR TREASURE," November 23–December 2, 2022.
- Vice-Rector's Office Exhibition Hall and Teruel Fine Arts Sculpture Workshop Hall, "TRASH OR TREASURE II," November 22–December 12, 2023.
- Barbastro-Monzón Diocesan Museum, Temporary Exhibition Hall: "Trash + Fantastic Beasts. Recycling as an Artistic Technique," from September 28 to December 19, 2024.

- Galeria Cozinha FBAUP: "Pictorial Deconstruction of Tiles," Porto, from February 24 to March 4, 2025.
- Lagrasse Abbey (France): exhibition at the Monstres & Chimères Congress. Images of Medieval Painted Ceilings, on May 30 and 31, 2025.

During its development, the research results have also been presented in various presentations at several conferences and events, such as:

- 2nd edition of the Regional Congress of Art and Emotion: EnamorArte of Education and Art, held in Teruel (March 15 and 16, 2019), with the presentation "The Beauty of Used Items" (in the experiences section).
- Extraordinary Course at the University of Zaragoza: Artistic Education in Contemporary Societies, held in Calatorao (Zaragoza, July 2 and 3, 2019) with the presentation entitled "Contemporary Art and Recycling in Schools".
- 1st International Conference on Arts, Society, and Education. Arts and Education in Contemporary Societies, held at the Center for Art and Natural Resources of Huesca (CDAN): Huesca City Council. With the presentation entitled "The Beauty of the Lived, of the Used: From Ready-Made to Trash Art".
- 13th Conference on Teaching Innovation and Educational Research of the University of Zaragoza, held in the Auditorium of the University of Zaragoza (September 5 and 6, 2019), with the presentation "The Beauty of the Used: Art and Recycling in Educational Environments (Trash Art)."
- European Researcher's Night at the Museum of Teruel (September 24, 2021), with the presentation "Trash Art as a Form of Reuse trash."
- 15th International Symposium on Mudejarism (October 6 and 7, 2021) with the presentation "Trash + Bestiary. Heritage as a source of inspiration."
- Nationale Nederlanden tour Plogging event (Teruel 2021).
- 5th Art Education Seminar. University of Zaragoza (February 18, Huesca 2022) with the presentation "Heritage as a source of inspiration. An example of a collaborative, participatory, and sustainable artistic project-process." The article was also published.
- Conference: Monsters & Chimères. Images of medieval painted ceilings. Medieval Abbey of Lagrasse. Les Arts de Lire Cultural Center.

Held on May 30 and 31, 2025, in Lagrasse (France), with the presentation/exhibition: "Trash + Medieval Bestiary: A Recycling Expo."
- USATIC Virtual Conference (Ubiquitous and Social: Learning with ICT). Zaragoza, June 24, 25, and 26. With the presentation: "TRASH + FANTASTIC BEASTS. Recycling and synchronous distance learning as a form of co-creation."

Conclusions

This project is designed to foster and produce collaborative gestures that provoke intersections between heritage, contemporary art, and recycling, utilizing new technologies that allow for synchronous teaching. It manages consumer habits and incorporates waste generated in the immediate surroundings as a didactic and artistic resource, with the intention of acquiring cultural value through reuse. The project reinterprets the bestiary of the Mudejar ceiling of Teruel Cathedral. It also provides the opportunity to interact synchronously with students from other countries, generating a joint co-creation project.

References

Agra, M. X. (2003). *La formación artística y sus lugares. Radiografía de la Educación Artística,* monográfico coordinado por Ricard Huerta. En Educación Artística. Revista de Investigación (EARI) n°1 (pp. 67–82), Institut de Creativitat i Innovacions Educatives. Universitat de València. https://dialnet.unirioja.es/ejemplar/337848

Borrás, G. M. 1999 La Techumbre Mudéjar de la Catedral de Teruel (n° 80–36). Caja de Ahorros de la Inmaculada de Aragón.

Brundtland Report. (1987, March 20). *Our Common Future*- ONU https://web.archive.org/web/20111003074433/http://worldinbalance.net/intagreements/1987-brundtland.php

Collados, A. & Rodrigo, J. (2015). *Retos y complejidades de las prácticas artísticas colaborativas y las pedagogías colectivas.* Pulso. Revista de educación, 38.

Demos, T. J. (2020). *Descolonizar la naturaleza. Arte contemporáneo y políticas de la ecología.* Akal.

Gálvez, E. (2013). Metodología activa favoreciendo los aprendizajes. En *Cuaderno de Apoyo Didáctico. Colección Construyendo juntos.*: Editorial Santillana S.A.

Kaprow, A. (2016). Entre el arte y la vida. In *Ensayos sobre el Happenig. Colección Héroes modernos.* Alpha Decay.

Moszynska, A. (2013). *SCULPTURE NOW.* Thames & Hudson world of art.

Prieto, J., & Ruiz, V. (2020). Desarrollando un bestiario sostenible. Arte y Memoria V, Teruel, Fundación Universitaria Antonio Gargallo (FUAG), Universidad de Zaragoza, pp. 75–110. https://zaguan.unizar.es/record/150295/files/BOOK-2025-279.pdf

Rancière, J. (2010). *El espectador emancipado*. Ediciones Manantial SRL.

Royal Academy of the Arts, London. (2000). en absolutearts.com. World Wide Arts Resources https://www.absolutearts.com/artsnews/2000/09/28/27501.html

Ruhrberg, K., Schneckenburger, M., Fricke, C., & Honnef, K. (2005). *Escultura* en Ingo F. In A. Walther (Ed.), *Arte del siglo XX (volumen II)*. Taschen.

Santos, F. (2021). *"Trash + Bestiario". Prácticas artísticas colaborativas, conciencia ecológica y valor patrimonial*. En J. Prieto (Ed.), "*Trash + Medieval bestiary. El patrimonio como fuente de inspiración*". Catálogo de la exposición en el Claustro del Museo de Arte Sacro de Teruel del 6 al 15 de octubre de 2021. https://zaguan.unizar.es/récord/148677

Tena Suck, A., & Rivas-Torres, R. (2005). *Manual de investigación documental. Elaboración de tesinas. Tercera edición*. Plaza y Valdez.

CHAPTER 13

Conclusions

Anna Maria Biedermann
and Francisco Javier Galán Pérez

The journey through the chapters of this book reveals a complex and multi-layered picture of the relationship between culture and sustainability. We have moved through rural landscapes and urban streets, tangible heritage and ephemeral digital networks, preserved pasts and imagined futures. Although each contribution has its own focus, methodology, and disciplinary language, all converge on a shared conviction: culture is not an optional part of sustainability but a constitutive dimension of it, as indispensable as environmental, economic, and social dimensions are.

A. M. Biedermann (✉)
Department of Design and Manufacturing Engineering, University of Zaragoza, Zaragoza, Spain

OAAEP Research Group, Funded by the Government of Aragón with FEDER Funds, Zaragoza, Spain
e-mail: anna@unizar.es

F. J. Galán Pérez
Predepartmental Unit of Fine Arts, University of Zaragoza, Zaragoza, Spain

OAAEP Research Group, Funded by the Government of Aragón with FEDER Funds, Zaragoza, Spain
e-mail: jgalan@unizar.es

A. M. Biedermann, F. J. Galán Pérez (eds.), *Interconnections between Culture and Sustainability*,
https://doi.org/10.1007/978-3-032-21061-6_13

One of the strongest threads running through the volume is the repositioning of culture from a peripheral concern to a foundational one. Historically, environmental, economic, and social dimensions have dominated sustainability discourse, supported by systems of measurement and policy frameworks. Culture, when acknowledged, has often been treated as an accessory or symbolic element. Yet the chapters demonstrate that culture is not an afterthought but the interpretive framework through which communities understand the environment, define economic priorities, and negotiate social relations. It motivates environmental stewardship, shapes governance processes, and gives meaning to technological or economic transitions. The lecture requires the reimagining sustainability itself as a cultural process, in which every other dimension is interpreted and enacted through shared values, narratives, and identities.

This centrality of culture brings us to one of the most persistent challenges identified by the authors: the difficulty of assessing cultural impact without reducing it to simplistic metrics. What gets measured tends to be valued in policymaking and funding decisions, yet the most significant aspects of culture often resist quantification. Several contributions in the book respond to this dilemma by proposing hybrid evaluation methods that combine quantitative indicators with qualitative narratives. The life cycle approach to exhibition design extends evaluation beyond the moment of public engagement to include creation and end-of-life phases. Methodologies for rural project design integrate participatory processes to ensure that indicators are locally meaningful. Heritage conservation strategies propose predictive and preventive evaluations that consider long--term social and cultural consequences, not just immediate physical preservation. Across these approaches emerges a shared principle: meaningful cultural evaluation must be situated, adaptable, and inclusive, grounded in the specific context yet informed by broader sustainability frameworks.

The present book also makes clear that sustainability is as much about sustaining relationships as it is about preserving objects or systems. Case studies involving heritage, memory institutions, and rural revitalization show that place-based culture is both a resource for and an outcome of sustainability. The Museo della Memoria illustrates how local institutions can bridge personal identity and global historical narratives, fostering bonds across generations and borders. The Fuendetodos case demonstrates how heritage linked to creative programming can strengthen cohesion, attract visitors, and inspire environmental responsibility. These examples underscore that heritage policies should not be static acts of preservation but dynamic strategies for resilience, engaging communities in ongoing reinterpretation and dialogue.

Culture, however, does not live only in monumental sites or organized events; it is embedded in the everyday spaces where people shop, work, and interact. The analysis of retail environments reframes commercial spaces as cultural and social hubs with significant implications for inclusivity and gender equity, challenging the idea that commerce and culture exist in opposition. Collaborative art projects such as Trash + Fantastic Beasts show how cultural creation can emerge from ordinary or discarded materials, transforming them into vehicles of shared meaning and environmental reflection. Both perspectives highlight culture's capacity to recode the everyday, creating new relationships with resources and waste.

In the digital realm, the discussion of technological ecosystems reveals that those infrastructures are not neutral; they are cultural environments that actively shape norms, values, and our sense of self. The transformation of products into network nodes and users into interconnected actors redefines how we inhabit both physical and virtual worlds. These systems can democratize cultural access and foster innovation, but they also risk homogenization, exclusion, and the erosion of local specificity. Designing technological systems with cultural sustainability in mind becomes an urgent task; one that demands diversity, contextual sensitivity, and user agency.

The story of the Splash sculpture offers a sobering reminder of the fragility of cultural narratives. Even celebrated works can lose relevance and visibility if they are not continuously integrated into public life. The sculpture's journey from symbolic centerpiece to obscurity encapsulates a wider truth: sustainability is not achieved simply by preserving an object but by maintaining the social and symbolic contexts that keep it meaningful. This insight resonates across the book: sustainability, whether of heritage, community identity, or technological culture, is a process, not a fixed state. It requires ongoing activation, reinterpretation, and reconnection.

A recurring element in many chapters is the centrality of participation. Whether through co-design workshops in rural planning, community involvement in museum governance, or collaborative art-making from recycled materials, participation emerges as both a cultural value and a practical sustainability strategy. Participation here is more than consultation; it is the co-creation of meaning, which fosters a sense of ownership and long-term commitment. This is crucial for sustainability, which depends on behavioral change and shared responsibility. At the same time, the book cautions against tokenism, showing that authentic participation must be carefully designed to respect local knowledge while linking it to broader objectives.

The diversity of disciplines represented such as design, fine arts, engineering, sociology or heritage studies, underscores that cultural sustainability cannot be addressed from a single vantage point. Interdisciplinary collaboration enriches understanding and methodology. This interdisciplinarity is not without challenges: it requires translation between vocabularies, negotiation of epistemological priorities, and often longer timelines for project development. The chapters show that such collaboration yields insights that could not emerge from disciplinary isolation.

Although the book is rich in specific case studies, their implications reach far beyond their immediate contexts. The life cycle approach to cultural products can inform other sectors; participatory rural methodologies could be adapted for urban regeneration; relational impact models for museums could be applied to libraries or archives; critical analysis of retail environments could be extended to other public spaces. Treating each case as both a local intervention and a methodological laboratory enables the construction of more flexible and transferable frameworks for cultural sustainability.

From an editorial perspective, one of the clearest messages of this work is that cultural sustainability must be named and claimed within policy agendas. It needs dedicated resources, appropriate evaluation methods, and professionals equipped to navigate both cultural and sustainability concerns. It calls for partnerships that cross institutional boundaries, linking cultural actors with environmental organizations, educators, and technology developers. And it demands patience, because cultural change, like ecological change, happens over the long term.

At the same time, this book has left us with questions that cannot be fully answered here: How can we better integrate cultural and environmental considerations into decision-making tools? How can we protect diversity in an increasingly interconnected digital world? How can we ensure that participatory processes are both authentic and scalable? These questions point to future research and continued experimentation, and they invite readers, whether practitioners, policymakers, or scholars, to see themselves as part of that work.

If there is one idea we hope will stay with you after closing this book, it is this: sustainability is, at its heart, a cultural project. Environmental stewardship, social justice, and economic equity cannot succeed without shared values, meaningful narratives, and collective practices that give them life.

As editors, preparing this volume has been more than an academic exercise. It has been an opportunity to listen, to learn, and to weave together perspectives that are as different as they are complementary. We hope that reading it has offered you the same: a chance to reflect on your own relationship with culture, to see new possibilities for action, and to imagine futures in which culture and sustainability are inseparable. The journey does not end here, but perhaps it continues with greater clarity, purpose, and connection. And in that ongoing journey, may culture remain both the compass that guides us and the terrain we inhabit.

Index[1]

[1] Note: Page numbers followed by 'n' refer to notes.

A. M. Biedermann, F. J. Galán Pérez (eds.), *Interconnections between Culture and Sustainability*,
https://doi.org/10.1007/978-3-032-21061-6

The manufacturer's authorised representative in the EU is Springer
Nature Customer Service Centre GmbH, Europaplatz 3, 69115 Heidelberg,
Germany. If you have any concerns regarding our products, please
contact ProductSafety@springernature.com

Printed and bound by CPI Group (UK) Ltd, Croydon, CR0 4YY
17/09/2026
02218454-0007